THOSE GASOLINE LINES AND HOW THEY GOT THERE

H. A. MERKLEIN AND
WILLIAM P. MURCHISON, JR.

A Fisher Institute Publication

ABOUT THE AUTHORS

HELMUT A. MERKLEIN has a B.S. degree in petroleum engineering from the University of Oklahoma, an MBA from the University of Dallas, and a Ph.D. in Economics from Texas A&M University. Fluent in several languages, he has authored three books on energy economics and macroeconomics, along with more than 100 articles in the field of energy and international economics. Dr. Merklein is dean of the Graduate School of Management at the University of Dallas. His experience includes almost ten years in the oil industry in petroleum production, reservoir engineering, and consulting.

WILLIAM P. MURCHISON, JR., received his MA degree in history from Stanford University, and has been a writer for Texas newspapers since 1964. He has been on the staff of *The Dallas Morning News* since 1973 as an editorial writer and columnist. Murchison has been a frequent contributor to national magazines and publications, and has a syndicated newspaper column. His research studies and subsequent editorial columns in the field of energy have been featured for the past several years on the editorial pages of *The Dallas Morning News*.

TABLE OF CONTENTS

INTRODUCTION

The thesis of this book is a simple one. It is that federal price controls and regulations caused the energy shortage.

Of course, energy is among the most complex of contemporary subjects. It is not our intention to deal with the full range of energy problems — with nuclear plants, with solar power, with gasohol and biomass — or, for that matter, with the vast range of conservation schemes, from mass transit to walking. Our concern, rather, is with oil and natural gas, the staple fuels, along with coal, of the world's richest, most technically advanced nation.

The fuel oil age has, in truth, been a relatively short one. It began with the discovery at Spindletop in 1901. There is no reason that it should terminate while there remain vast deposits of oil and gas still to be discovered and produced in the United States.

The roadblocks that government policy has thrown in the way of discovery and production have been formidable and perverse. By pressing down on energy prices for a quarter of a century, the government has greatly discouraged full utilization of the oil industry's dynamism and knowhow. Milton Friedman, Nobel laureate in economics, has pointedly asked: "Do you want to produce a shortage of any product? Simply have government fix and enforce a maximum price on the product which is less than the price that would otherwise prevail." So it has been with oil and gas. Try as the government might to fix prices at a level satisfactory to producers and consumers alike, it failed utterly. In the nature of things, the government cannot know what price is right; only the free marketplace can know.

Distrusting the marketplace — at least as it pertained to energy — the government hamstrung its natural operations. As the dreary history of economic controls might have suggested, such an attempt was bound to miscarry, bringing upon us the shortages that vex us now and will vex us for years to come.

The wonder, perhaps, is that the failure of our government's energy policy should not be more widely noted or acknowledged. The shortages, to be sure, are so great as to be common knowledge — indeed, part of everyday experience. And yet the cause is not frequently enough recognized. Instead the OPEC cartel is denounced, or the oil companies railed at. As if either had sought to price oil so low as to discourage exploration for new supplies and to encourage waste. Now, when the nation remains unsure as to what energy policy it wants — one embracing more controls, or fewer, or even none at all — telling the real story of the energy crisis seems to us a matter of some urgency. This is that story.

H. A. MERKLEIN
WILLIAM P. MURCHISON, JR.

Dallas, April, 1980

CHAPTER 1

THE DEVELOPING SHORTAGE

The origin of a nation's troubles, traumas, vexations is notoriously hard to date. Which makes the energy crisis a rare commodity. Its start can be *precisely* dated, for it was on June 7, 1954, that the United States Supreme Court handed down its decision in the case of Phillips Petroleum Co. vs. Wisconsin. A quarter of a century later, the nation continues to pay the bill.

The Phillips decision — so it is generally called — arose from an attempt by the Phillips Petroleum Co. to negotiate a higher price for natural gas it sold its customers in Milwaukee. From three cents per thousand cubic feet (mcf), the company proposed jumping the price to four cents. Milwaukee sued to prevent the increase. And so at last the case reached the U.S. Supreme Court. The court declared that Congress had meant, in the Natural Gas Act of 1938, which regulated interstate pipelines, to bring under federal control the wellhead price paid natural gas producers operating in interstate commerce. This was odd, for the act had explicitly stated that "the provisions of this chapter . . . shall not apply . . . to the production or gathering of natural gas." It was not the first nor the last time the high court would find in the language of Congress a coloration invisible to the naked eye.

But if the Phillips decision was questionable law, it proved to be calamitous economics. Through it twine the roots of the energy crisis.

The notion of price controls, which the decision advanced, was no new one. Government attempts to fix the price of a commodity — or of all commodities — date back to the shadowiest centuries before Christ. Among the rasher attempts was that of the Roman Emperor Diocletian, whose edict of 301 A.D. froze the prices of most goods and services. Until, as the historian Lactantius records: "There was much blood shed upon very slight and trifling accounts; and the people brought provisions no more to markets, since they could not get a reasonable price for them, and this increased the dearth so much, that at last after many had died by it, the law itself was set aside."

Of economic controls, Robert Schuettinger and Eamonn F. Butler conclude in their recent book, *Forty Centuries of Wage and Price Controls*[1], "In all times and in all places they have . . . invariably failed to achieve their announced purposes."

It was to be expected by those with a rudimentary grasp of economics, not to say common sense, that the Phillips decision would bring down upon the producers and consumers of natural gas the same consequences that controls always have let loose. Controls would stimulate demand for gas, because the price would be set artificially low. Unless it were fixed below the price that free competition would command, why bother with controls at all?

But that was merely half the picture. The other half was that, even as demand rose for natural gas, the supply was guaranteed sooner or later to fall short. The merchants of Diocletian's day would easily have understood. Unable to command a fair price, one that rewarded their labor over and above the cost of production, they had "brought provisions no more to markets." Federally controlled prices for natural gas were certain to have much the same effect. A smaller return on invested capital meant diminished enthusiasm for risking more capital in the chancy and expensive business of exploring for natural gas.

Persuaded by gas producers' arguments to this effect, Congress in 1957 passed a bill stripping the Federal Power Commission of its newly acquired power to fix natural gas prices at the wellhead. Then disaster struck. An oil company lobbyist was accused by Senator Francis Case of South Dakota of offering a bribe in return for his support of the bill. Between the economic principle at stake and the unprincipled act of the lobbyist there was utterly no connection. But when the decontrol bill reached his desk, President Eisenhower in-

[1]The Heritage Foundation, Washington, D.C., p. 147.

dignantly vetoed it. However sincerely motivated, the veto was perhaps the single most disastrous action on energy that any president has taken — and there is much in this lamentable category from which to choose.

The continuation of controls — Congress having failed to override the veto — meant the eventual depletion of the nation's natural gas supplies. It was not the sole cause of the energy crisis. It was the trigger. Not that all saw it being squeezed. Since the 1940s, the nation had reveled in a surplus of natural gas. Twenty-year contacts written during the '30s and the '40s had specified prices as stupefyingly low — by standards of the late '70s — as 2 cents per mcf. Such were the benevolent consequences of competition. It was not the federal government that arranged such bargains, hammering out their terms, enforcing those terms with its full majesty and might. It was the free marketplace, harmonizing supply and demand, making sense of widely divergent needs — of consumers for gas, of producers for fair return on invested capital. A happy state of affairs indeed, but destined not to last.

The virus of the Phillips decision was slow in working its way into the national bloodstream. There were at first no distressing symptoms. The Federal Power Commission set about its appointed task of pricing interstate natural gas by figuring each producer's "cost of service." Hearings were set, conducted, recessed, begun anew. Most proved of nearly interminable length, like the celebrated case in Dickens, Jarndyce vs. Jarndyce, which had gone on so long that no one remembered how it started. By 1960, the FPC had settled a mere 11 rate cases, with more than 3200 backed up awaiting action.

Accordingly, the FPC decided to set prices on an area-wide basis. Not that matters were thereby greatly expedited. The Permian Basin case was opened in 1961; seven years later, the Supreme Court affirmed its terms. The Southern Louisiana cases, which commenced also in 1961, were not finally determined by the Supreme Court until June 10, 1974. Not the least effect of the Phillips decision was the uncertainty with which it left producers as to the price they would receive.

Equally baneful consequences were meantime at large. A surplus is like puppy love — joyous and pulse-racing while it lasts, but fated to last only a while. The sheer abundance implied in the word "surplus" depresses prices. If there is plenty — indeed more than enough — to go around, no desperate buyers are available to bid the price up. Yet the low prices that surpluses enforce put increasing pressure on supplies.

It is the ancient and infinitely human love of bargains that is at work. With prices low, a product — especially an essential product — is quickly snapped up. So it proved with gas. As it was, the low prices generated by the surplus increased demand. So did those low prices dictated and maintained by the Federal Power Commission. Throughout the '60s, according to one study, the regulated price of interstate natural gas was 6 cents per mcf cheaper than it would have been at the so-called "market-clearing" level — the level to which the overall U.S. supply and demand (both interstate and intrastate) would have drawn prices if left free to do so.

With gas relatively cheap, it only made sense to buy and burn it. Between 1946 and 1970, natural gas production more than quadrupled. Yet there was another side to this resplendent coin: proved reserves of gas less than doubled. The nation, in short, was consuming gas much faster than it was replacing it.

This took place for an altogether obvious reason: federal controls on gas prices had dampened the gas producers' incentive to drill the costly new wells that alone could maintain or increase domestic supplies. Gaswell drilling peaked in 1962 — the year after the FPC initiated the Permian and Southern Louisiana rate cases. From a total of 5,848 wells drilled that year, the number by 1970 had fallen below 4,000.

Yet a distinction has to be drawn. There was not, before 1978, one unified market for natural gas; there were two: the interstate market, where federal price controls prevailed, and the intrastate market, where the law of supply and demand held sway. If gas were discovered, say, in South Texas and sold in the Midwest, the Federal Power Commission set the price. But if the gas were sold in Houston, the government lacked jurisdiction. The seller and the buyer agreed as to what should be paid.

At that, the free market price was generally cheaper, at least until 1973, than the controlled one, into which had to be figured the legal and administrative costs that go hand in hand with government regulation. In 1973, Texas consumers were paying an average 19.5 cents per mcf for gas produced in their state. But had the gas crossed a state line, the price would have been 21.4 cents. Such were the blessings of regulation.

The development of a dual pricing system had especially potent effects on the intrastate market, where the absence of controls permitted a price response to the developing shortage of newly found natural gas. Starting in 1970, consumers began to bid up the price of such gas, whereupon producers increasingly allocated their resources to the exploration and development of the higher-priced intrastate gas. In 1968 and 1969, for example, only 35 per cent of the newly-found natural gas had been committed to intrastate markets. By 1970, intrastate gas commitments reached 69 per cent, and thereafter the figure would fluctuate (80 per cent in 1971, 53 per cent in 1972, 83 per cent in 1973, and so on). The precedent was established: henceforth the regulated interstate market would increasingly be starved for gas, while the free intrastate market would flourish. After 1973, when the energy crunch began in earnest and free-market prices rose sharply, the producers' attachment to the intrastate market grew invincible.

Price controls — so thoughtfully designed for the protection of consumers — were beginning to pinch and inconvenience those same consumers. Though, to be sure, very few even noticed at first. Throughout the '60s oil industry spokesmen tried valiantly to warn of the coming crunch in energy supplies. One of them, Michel T. Halbouty of Houston, was amazingly far-sighted. On Nov. 3, 1960, Halbouty said in a Los Angeles speech that, "Unless there is an appreciable and sustained turnaround in our exploratory activities, I can safely predict that between now and 1975 we will have an energy crisis in this country which will cause repercussions throughout the width and breadth of this great nation of ours like a devastating earthquake."

To many, it must have seemed a strange time for such apocalyptic talk. In 1960, the world abounded in crude oil — more crude oil, in fact, than it could consume. This was what distressed Halbouty, a veteran wildcatter and geologist. The surplus, he saw, was depressing the search for new supplies. Whereas in 1956, a record 58,000 domestic wells had been drilled, only 46,000 were sunk in 1960. Moreover, Halbouty recognized that as demand increased, so the world's surplus producing capacity would fall. Great new reservoirs of oil would have to be discovered if demand was not soon to outstrip supply. "Some of these days," he said, "the shortage will catch up with us, and then the people will say, 'The industry is to blame, why weren't we told?' Well, I am telling them now."

Yet the '60s, as a decade, was an inpropitious time to be raising new specters of crisis before the American people's eyes. They had enough specters as it was. Vietnam, civil rights, campus revolt, long hair, the drug scene: it was as much as any civilized nation could reasonably be expected to cope with at a given moment. The warnings of an energy crunch could scarcely be heard above the thunder proceeding from Southeast Asia.

By 1971, the war, in the terminology of the Nixon administration, was "winding down." There was more opportunity to notice that the nation's energy problems were winding themselves tighter and tighter with each passing month. Domestic crude oil drilling had been in decline since 1956; gas well drilling followed suit after 1962. In 1968, the nation for the first time consumed more gas than it added to its reserves. Such hints of impending trouble made themselves known, if only in circles intimately concerned with matters like these. In 1970, the White House created an interagency Oil Policy Committee charged with stabilizing oil prices and maximizing supplies. The following year, the Office of Emergency Preparedness, in a 70-page report to President Nixon, warned that, "In the long run, future crude oil price increases may be necessary if fuel supplies are to be developed consistent with our national security requirements." The OEP noted that since 1959, when import controls were established by Congress, crude oil prices had risen less than the wholesale prices of all other commodities — a clear indication that oil prices were lagging behind inflation. Likewise, the report fretted that low gas prices afforded the natural gas industry too little incentive to drill exploratory wells.

By March of 1972, Gen. George A. Lincoln, director of OEP, was warning a Senate committee that, "There is, on the near horizon, a shortage of gas which will cause curtailment . . . In effect, we are moving toward a gas-rationing system." The same day, Columbia Gas Transmission Co. announced it was unable to increase wholesale gas sales to its customers, who stretched from Southwest Ohio to Northern Virginia. The Washington Gas Light Co. said it would no longer connect up new customers, there being insufficient gas to supply them. By the British Petroleum Co.'s estimate, the free world's production of oil would fall short of its consumption by 600,000 barrels a day. The situation was very much as *Fortune* magazine described it that fall: "In the sense that our past usage of energy resources has been reckless, the 'joyride' — as some have called it — is finally coming to an end."

Even the Federal Power Commission had been coming to understand that things were amiss. John N. Nassikas, the new chairman appointed by President Nixon in 1969, had hardly settled in office before he called for a "Nassikas round" of price increases. The FPC duly raised area-wide price ceilings, exempted small producers from the ceilings, and authorized producers making emergency sales to breach the ceilings for periods up to 180 days. Likewise, the commission established an "optional pricing" system; producers might negotiate new rates, subject to commission approval. It was the right kind of stimulus: gas well drilling began to pick up. Yet another FPC action established even more plainly the failure of price controls. In 1972, the commission approved the importation of Algerian liquefied natural gas (LNG) at 91 cents per mcf, compared with new-contract prices of 26 cents in the interstate market and 40 cents in Texas and Louisiana. Unable to persuade Americans to produce gas at bargain-basement prices, the United States would buy gas three times as expensive from a nation traditionally unfriendly to the West. There was a perverse kind of beauty to it, a fascinating disharmony of intent and effect. Was the meaning of consumer price-protection fully revealed at last? And was that meaning just this — that the free market, not government, affords the only economic protection worth having?

The FPC's gestures to the producers were too little — and far, far too late. In January, 1973, the nation suddenly became aware that there was too little oil and natural gas to go around. In Denver, high schools went on a three-day week. Factories in West Virginia, Illinois, and Mississippi closed down for lack of fuel. Grain shipments were stranded on barges on the Mississippi and Ohio rivers. That month, for the first time, energy became major news, even making the cover of *U.S. News and World Report*.

It was hardly a surprise to those who followed trends within the petroleum industry. Natural gas was running low for reasons already alluded to. Fuel oil was in short supply for similar reasons. On Aug. 15, 1971, hoping to break the momentum of inflation, President Nixon had imposed a wage and price freeze on the whole economy. The freeze found fuel oil prices at their seasonal low, the demand for heating oil in August being less than brisk. Conversely, gasoline prices were at their seasonal high, what with the "driving season" still in full swing. The upshot was that refiners had little incentive to produce more fuel oil; it repaid the investment less than did gasoline. The situation was scarcely helped by new environmental laws requiring the installation of equipment on automobile engines that reduced air pollution but increased gasoline consumption. It meant the refiners had to make more gasoline than before just to stay even. And so, by November, 1972, the nation's fuel oil reserves were three million barrels below the level of 1971.

The winter disappeared at last, swallowed up by warmth and greenness, but not the talk of an impending emergency.

The Texas Railroad Commission, the state agency that oversees the Texas oil and gas industry, ordered in March all-out production of oil (consistent with safeguards against too-quick depletion of a reservoir). Yet notwithstanding the industry's attention to gasoline production in the previous months, the spring of 1973 brought shortages and rising prices. In March, April, and May, gasoline prices climbed at a 13 per cent annual rate.

On April 18, 1973, President Nixon sent to Congress the first of three energy messages that would come before year's end. The President had last addressed the nation on energy in 1971. Then, the emphasis was on how to satisfy the "accelerated growth in demand" for clean, non-polluting fuels — like natural gas. Yet there had not been any indication of a need for price decontrol. Now, with energy firmly implanted at the forefront of the nation's concerns, the President sought to correct his omission. He called for deregulation of the price of gas from new wells, with the Interior secretary authorized to impose a ceiling price should the need arise. "It is clear," said Nixon, "that the price paid to producers for natural gas in interstate trade must increase if there is to be needed incentive for increasing supply and reducing inefficient usage."

The President's call was seconded by Federal Power Commission Chairman Nassikas, who released a study indicating that, from the end of 1966 to mid-1972, gas reserves had declined 26 per cent and were 9 per cent below the industry's estimate. Only when prices were allowed to float up to the market level, said Nassikas, the reluctant regulator, would the gas shortage be resolved.

John A. Love, director of the White House Office of Energy Policy, undertook to show just how underpriced was natural gas. If crude oil were sold at the equivalent gas price of 22 cents per thousand cubic feet, domestic crude would sell for 72 cents per barrel, compared to $1.00 a barrel for foreign crude. So cheap was gas, Love complained, that industry was increasingly monopolizing its acquisition at a time when new homes were being denied connections.

Besides calling for price decontrol, Nixon urged more

offshore exploration, the building of deepwater ports for oil tankers, and more research and development funds for future energy sources. He also ended the 14-year-old mandatory oil import quota system, established in 1959 to restrict imports of cheaper foreign oil and thus, theoretically, increase demand for domestic oil. The President's commitment to a free market in energy was less than total, one must note. Just a month earlier, he had reimposed mandatory price controls on the country's 23 largest oil companies as part of "Phase 3" in his economic stabilization program. Earlier controls had recently expired, but the industry's respite was to be brief. With demand forcing up prices, the President directed that the companies hold their averge monthly price increases to 1 per cent. Crude oil, gasoline, and heating oil were unaffected.

By early summer of 1973, serious fuel shortages were developing around the country. For lack of fuel, gasoline stations were forced to shut down. "Already," said *U.S. News and World Report* in July, "the chances are about 50-50 that any service station you drive into will impose a limit on the amount of gasoline you buy, or will be closing earlier than usual to stretch out limited supplies." Black markets began to appear in some farm states.

On June 1, 1973, gasoline inventories were 6.7 per cent below year-earlier levels. Yet demand was 6 per cent higher. Nor was gasoline the only fuel in short supply. A survey by the National Federation of Independent Business in early summer showed 63 per cent of the fuel oil dealers in 23 states had experienced cutbacks in supply. Seventy-four per cent were paying higher wholesale prices. There were angry cries by politicians against the major oil companies, to whom was imputed the guilt for so surprising a state of affairs. But where, in truth, was the surprise? The shortages and the price increases of 1973 were no more than the logical outcome of trends set in motion years earlier, as foretold again and again by the petroleum industry. The government had tried to fix what it deemed a fair and appropriate price for a major commodity. And naturally it had failed, as governments always fail in such circumstances, the marketplace being oblivious to the finest calculations of the best-intentioned government regulators. Quite soon, even greater proof was forthcoming.

CHAPTER 2

THE EMBARGO AND ITS AFTERMATH

On May 5, 1973, four Arab nations — Libya, Algeria, Iraq, and Kuwait — briefly suspended the pumping of oil. It was a gesture — nothing more; a sign of solidarity in the struggle against Israel. A Western oilman in Libya dismissed the shutdown, saying, "Maybe a couple of guys took a break for an hour or two, but there has been no halt in shipments."

Indeed, the shutdown attracted little public notice, consigned as it generally was to the inside pages of the newspapers. But its meaning went far beyond short-term implications for supply and demand.

For years, the Arabs had talked of using their oil as a weapon against Israel's Western supporters. In 1967, during the Six-Day war, every Arab producer except Algeria had totally shut down its oil fields. But the United States, Venezuela, and Indonesia had responded by increasing production. Within a few weeks, the embargo was a shambles.

Half a dozen years had nevertheless made a significant difference in the West's ability to parry "the oil weapon." Since the late 1960s, supplies of both natural gas and crude oil had dropped dangerously. The events of the winter and spring of 1973 had demonstrated that the United States could not, out of its own resources, supply its own people, much less the peoples of allied Western nations. By September, 1973, the United States was importing every day a record 1.2 million barrels of Arab oil. Only the year before, the figure had been just 800,000 barrels daily. From all overseas sources, the U.S. was buying six million barrels a day — 32 per cent more than in 1972.

Not surprisingly, the mood of the Organization of Petroleum Exporting Countries, founded in 1960 to keep oil prices from dropping, turned ever more militant. In September of 1973, OPEC renewed its call for higher crude oil prices. Most of the Arab exporters announced that on October 8, in Vienna, they would begin dickering with the oil companies for complete revision of the 1971 Tehran agreement, whereunder oil prices were strictly fixed.

As it happened, world events eclipsed the Vienna meeting even before it could begin. On October 6, the Jewish Day of Atonement (Yom Kippur), Syrian warplanes attacked Israeli positions in the Golan Heights; simultaneously, Egyptian troops were thrown at Israeli forces along the Suez Canal. For the third time since 1948, the Middle East was aflame.

The Arab armies fared poorly after some initial successes. Not so the Arab oil ministers. Eleven days after the war had begun, 11 Arab nations announced that they meant to bludgeon the United States into a change of Middle Eastern policy. They would do this by reducing oil exports 5 per cent a month until the Israelis withdrew from Arab territories occupied in 1967 and pledged respect for the rights of Palestinian refugees. On October 18, Saudi Arabia, which exported nearly 800,000 barrels of oil daily to the United States, threatened to suspend these shipments unless the U.S. recanted its support of Israel. The United Arab Emirates cut off the 100,000 or so barrels daily it was sending to America. Libya followed suit on the 19th. By October 21 every drop of Arab oil destined for America, and for other nations thought sympathetic to Israel, had been shut off.

Europe, which imported more than half its oil from the Middle East, was clearly in for hard times; likewise Japan, 45 per cent of whose oil came from the Arab countries. But even the United States — the energy-rich United States, where the fuel oil age had begun! — had cause to be fearful. The lost oil could not be replaced. Texas, the country's principal oil state, had for 20 months been pumping its wells at maximum capacity.

The energy shortage had been serious enough the previous spring and summer. Now the shortfall between supply and demand was projected at 2 million barrels a day — 12 per cent of the nation's crude-oil needs.

A battered and harried President Nixon — "the deplorable Watergate matter," as he called it, was raging unabated — appeared on television November 7 to announce that "We are heading toward the most acute shortages of energy since World War II." But the President's response was curious. If supplies were short, surely right reason indicated the adoption of measures to increase supply. The President instead proposed measures to spread the shortage more evenly. Americans were asked to use "less heat, less electricity, less gasoline." By executive order, heating oil allocations were cut and temperatures in federal offices reduced to 65-68 degrees Fahrenheit. The Congress was urged to institute year-round daylight saving time, lower the speed limit, relax environmental restrictions, and "impose special energy conservation measures." Of price decontrol for oil and natural gas there was, on this occasion, no mention. The presidential message was weighted almost wholly toward conservation.

This was curious not only because Nixon had previously endorsed decontrol of gas prices, but also because higher prices for oil and gas already were boosting the search for both commodities. "The outlook for increased U.S. oil and gas drilling," a spokesman for drilling contractors had said in October, "is brighter than at any time during the past 17 years." The September, 1973, average of 1,270 drilling rigs at work was the highest since December, 1969, and 10.5 per cent above the level of September, 1972. It was fully 28.5 per cent higher than the level of the previous April. "As expected," said Warren L. Baker of the International Association of Drilling Contractors, "rising U.S. crude oil and natural gas prices are having the desired effect of stimulating the search for new production."

By October 25, fighting in the Middle East had ended. The combatants eyed each other warily across tenuous cease-fire lines. But the embargo against the United States remained in force. Indeed, there was worse to come: on December 23, Iran, Iraq, Saudi Arabia, Kuwait, the Emirates and Qatar doubled the asking price of their oil — from $5.11

a barrel to $11.65. Five days later, Venezuela jumped its price from $7.74 a barrel to $14.08. Similarly large increases were posted on December 31 by Libya, Nigeria, Bolivia and Indonesia. So oil henceforth would not just be hard to get; it would be painfully expensive.

The old year ended, the new began, with Americans lined up for the privilege of buying gasoline at prices 35 per cent higher than the year before; turning down thermostat dials to conserve precious heat; contemplating the darkened skylines of the great cities, where nonessential lighting had been extinguished. Gasoline rationing, the most stringent and fearsome of conservation measures, was never invoked. But the Bureau of Printing and Engraving printed ration coupons just in case. The nostalgia of "Auld Lang Syne," as it was played to welcome the new year, had a keener edge than usual. The old days, the beloved old days of cheap and abundant energy, lay in the past. A cup of kindness might appropriately be drunk to them, for they were gone beyond recall.

It was plainly time for the nation to get down to the urgent business of adding to its still considerable trove of energy. But for so urgent a task neither the Congress nor the President had much appetite. As in late 1973, the talk throughout 1974 was mainly of conservation. It was talk made easier, perhaps, by the lifting of the Arab embargo on March 18, after which gasoline and fuel oil once more became plentiful.

For most of 1974, Watergate, climaxed by the resignation of President Nixon in August, engaged the nation's attention. It befell Nixon's successor, Gerald Ford, to try and achieve what his predecessors had failed to undertake — the liberation of the oil and gas market.

On January 23, 1975, Ford declared that on the following April 1 he would lift controls on "old" domestic oil — that is, oil found before 1973. The federally controlled price for "old" oil was only $5.05 a barrel, less than half the world price. ("New" oil, found during or after 1973, sold near the uncontrolled world price.) Just a week after this message Ford asked Congress to deregulate the price of natural gas.

The Federal Power Commission had, in truth, been

moving towards decontrol — nearer, at any rate, than ever before. On June 21, 1974, it had set a national rate of 42 cents per thousand cubic feet — admittedly well below the free-market intrastate price of almost $2, but substantially better than interstate producers had ever before enjoyed. What was more, an escalator clause would permit prices to increase 1 per cent a year. In early December, the FPC raised the base price again — this time to 50 cents per thousand cubic feet.

Yet elsewhere in Washington, D.C., decontrol was widely regarded as the work of the devil. The Library of Congress, a supposedly non-partisan agency pressed into ideological service for the occasion, wrote a report for Congressman John Moss speculating that "price incentive alone may have very limited impact on gas production." The library cited the failure of oil production to rise after the big price increase, ignoring the sharply escalated rate of drilling that higher prices had produced — to say nothing of the effects of price incentives on the unregulated intrastate gas market. Officially, Congress' Democratic leadership, at loggerheads anyway with the Republican President, found it unjust that OPEC should be allowed to set the world price of energy; decontrol, the leadership reasoned, would only aid and abet the Arabs.

Consequently, Ford shifted strategy. Rather than formally submitting his program for immediate oil-price decontrol, he waited until July to ask for a 30-month phase-out of controls, during which period price ceilings would be maintained. A long wrangle resulted, lasting the rest of 1975. In the end, Congress insisted on extending mandatory controls into early 1979 and on lowering, what was more, the average per-barrel price of domestic oil. This average price, figuring in "old" and "new" oil, had been $8.63 a barrel. For the short run Congress rolled the average back to $7.87, which meant that the price of oil discovered between 1973 and 1975 had to be reduced for the sake of maintaining the new, lower average.

The President had hoped for major progress on price decontrol. Instead, he received a setback. Though it was agreed that the average price should rise gradually over the 40-month period before the authority to control crude-oil prices expired, oil producers had met with a nasty shock. They had thought they were operating at least partly in a free market situation, where newly-found oil could be priced at the world level, even if oil previously discovered sold only at the archaic price of $5.05 a barrel. Congress now, far from consenting to total decontrol, was *tightening* the controls process. It was hardly an incentive to risk valuable capital in a search that, whatever the excellence of modern geophysical and geological techniques, depended in no small way on good luck. In 1976, oil-well drilling rose only slightly above the 1975 level.

Only a little more progress was made on gas price decontrol. There had been predictions of a cold winter, during which demand for heat would rise sharply and gas shortages would probably occur. The Senate was at last moved to a long overdue consideration of free market economics. It passed the Pearson-Bentsen bill, deregulating the price of new onshore natural gas as of April 4, 1976; new offshore gas would be freed from controls on January 1, 1981. The House, less enthusiastic about competitive enterprise, dragged its feet. Meanwhile, the FPC rescued the lower chamber from the necessity of taking a stand one way or the other. On July 27, 1976, the FPC nearly tripled the ceiling price for "new" gas — from 52 cents per mcf to $1.42. Gas discovered in 1973 and 1974 was permitted to sell at $1.10. With new intrastate gas selling at $1.50 to $2.00 per mcf, the price gap between the two markets was narrowing significantly.

The FPC reasoned that gas producers were bearing higher taxes and drilling costs and therefore deserved a higher rate of return on invested capital. It was an implicit admission, not just of present needs, but of the inherent shortcomings of price regulation. However, the House failed to carry the FPC's logic to its natural conclusion, which was complete decontrol of gas prices. Self-anointed "consumer" spokesmen were angry enough about the FPC; it was better, the House leadership reasoned, to let sleeping dogs lie. And so decontrol was put quietly to rest.

The 1976 presidential election brought Jimmy Carter of Georgia to the White House. The nation's energy dilemmas had in the years since 1973 become more urgent than ever. The price of domestic oil — an average $8.19 per barrel by 1976 — had failed to inhibit consumption. The daily production of domestic oil had fallen by one million barrels since 1973. Imports had risen by the same amount. Worse, these imports were increasingly expensive. The landed cost of Arab oil in late 1976 was $13.40 a barrel, four times the pre-embargo level. At the same time, the United States was compelled to purchase more and more of it as Canada, a once-dependable supplier, had announced the phasing-out of exports to America by 1981. The Federal Energy Administration had earlier forecast that oil imports could, by 1985, be cut to between three and five million barrels a day (from seven million in 1976). Now it was obliged to doubt that imports could possibly fall lower than 5.9 million barrels. Besides an active program of conservation, tailored to avoid waste of energy, the United States plainly needed a program to bolster the supplies it still possessed.

During the presidential campaign, candidate Carter had indicated his support of price decontrol for natural gas. Yet somehow, between election day and the occasion of his first energy message (April 20, 1977), he changed his mind. The thrust of the Carter energy package was conservation. The President would raise oil prices all right: he would raise them through taxes. The extra money would not therefore go to the oil companies for drilling and exploration programs, but rather to the federal government for partial redistribution to the taxpayers. As for natural gas, its price would be tied to the higher prices fetched by crude oil. But federal price controls would endure. They would *more* than endure: they would be *extended* for the first time to the intrastate gas market where unregulated prices, borne higher and higher by the tides of demand, had kept the market plentifully supplied. The President advanced other conservation ideas, the most hotly debated of which was a 5-cent-a-gallon tax on gasoline. The message, however, was devoid of proposals to increase production, except insofar as he considered a firmly controlled, nationwide natural gas price of about $1.75 per mcf a sufficient incentive to drill more wells. The Carter program was attacked by congressmen, editorial writers, producers and consumers alike; for the most part, it got nowhere during 1977. Neither did countervailing proposals to deregulate oil and gas. Nor did the nation itself get anywhere during 1977.

It was 1978 — a year in which energy consumption grew by 2 per cent, with oil imports rising to 45 per cent of the total consumed — before action was taken on energy. And even then, the action taken troubled those who had insisted, in season and out, that the free market must be given latitude to work.

On October 15, 1978, at 7:30 a.m., the President's 18-month-old energy program at last won congressional enactment. Or, rather, a few surviving portions of it won enactment, for the program was much altered from its original conception. The use of coal in future utility plants was mandated; new decorative outdoor lights were outlawed; supposed energy-saving methods of utility pricing were recommended to the states; tax credits were voted for the installation of energy-saving devices; heavy taxes were ordered for "gas-guzzling" automobiles.

For all its deficiencies, the Congress' (for it really by then had ceased to be the President's) energy program at least set the natural gas industry walking stiff-legged toward decontrol. As the President had proposed, price controls were extended to the intrastate gas market. The national price for "new" gas was to be $1.75 per million British thermal units (MMBtu), roughly one mcf, retroactive to April 20, 1977; prices would rise monthly to cover inflation, and other upward adjustments were mandated. Congress specified that price controls would expire on January 1, 1985, on new gas, deep new onshore wells, and existing intrastate contracts priced at more than $1 per MMBtu on December 31, 1984.

It was decontrol of the most grudging sort. The natural gas producer found himself hemmed in by all manner of special provisions. "New" gas meant gas from wells drilled at least 2.5 miles from, and at least 1,000 feet deeper than, existing wells. Emphatically it did not mean gas "withheld" from production, gas from Prudhoe Bay in Alaska, or gas from untapped reservoirs that adjoined new wells being drilled. In numerous other ways Congress had underscored its reluctance to acknowledge the workings of the laws of economics, as men had been witnessing to them for two centuries. It formerly was enough that a producer had discovered gas. What mattered now was *when, where* and *how* he discovered it. One thing, at any rate, came of the 18-month-long fracas: The principle of ultimate decontrol was at long last established. Faced with the choice of further entrenching a 24-year-old mistake or of moving toward freedom, Congress chose the latter course. The grudging nature of its decision cannot conceal entirely that a break had been made with history and ideology.

This, to be sure, left the question of oil prices unresolved. The 1975 extension of price controls was due to expire in September, 1979. A decision was consequently forced upon the President, who resolved it by announcing on April 5 that he would phase out oil price controls by September 30, 1981.

"Federal government price controls," said the President, "now hold down our own production and encourage waste and increasing dependence on foreign oil." It was a view that economists, oil producers and presidents had been urging for years. At last it was to prevail.

Yet what was this? The President had not finished speaking his piece. Yes, controls were on the way out. But the higher revenues that decontrol would produce had to be diverted from the oilmen. "Unless we tax the oil companies, they will reap huge and undeserved windfall profits. We must impose a windfall profits tax on the oil companies to capture part of this money for the American people."

So that was the size of it: higher prices for oil but, for the producers of oil, only a portion of the money those prices would bring in. Profits that might otherwise return to the ground in the continuing search for one of the world's most treasured commodities, energy, would cascade into the U.S. Treasury, where not a single drop of oil, so far as anyone knows, has ever been found.

The President's words reached an America increasingly nervous about its future. Gasoline lines once more were forming. For on December 26, 1978, Iran, the world's fourth largest oil producer, had ceased to produce oil. The Ayatollah Ruhollah Khomeini's Moslem revolution had paralyzed the oil industry. Memories of 1973-74 took on flesh as gasoline lines lengthened that spring and summer, and prices rose in the wake of yet another OPEC price increase. It was clearer than ever before that federal energy policy, its foundation laid down by the U.S. Supreme Court in 1954, was a monument to failure and futility. Yet the foundations, though crumbling, remained. The most grievous mistakes, it would appear, are the hardest to rectify.

CHAPTER 3

THE NATURAL GAS SHORTAGE

Federal control of natural gas prices began with the Phillips decision of 1954, in which the United States Supreme Court instructed the Federal Power Commission to begin regulating the wellhead price of gas sold across state lines.

The story for a time thereafter is a bland one — a tale of rising reserves and moderately stable prices. But only for a time. The FPC's attempt to fix prices led eventually to discomfort, then hardship; at last, to the brink of calamity.

The value of natural gas as a *fuel* had long been recognized. By 1897, gas was so used in 12 states, particularly Pennsylvania and Indiana — though two years later it supplied only 3.2 per cent of the nation's energy. The attractions of gas were obvious: it was produced as a byproduct of oil; it burned cleanly; best of all, perhaps, it was cheap, owing to its abundance and to the ease of producing and transporting it.

In 1954, the year of the Phillips decision, the nation produced 9.4 trillion cubic feet of natural gas. Its proved reserves — gas still below ground but definitely obtainable — totaled 210 trillion cubic feet, enough to last 22.5 years. Here was a natural resource valuable in the highest degree — efficient, inexpensive and plentiful.

The equation was dramatically changed, however, by the Supreme Court's finding that Congress, in the Natural Gas Act of 1938, had meant to control the wellhead price of interstate gas. The entire purpose of controls is, of course, to suppress prices and so to "protect" the public from predatory capitalists. The controlled price of natural gas was low enough throughout the first two decades after the Phillips decision: an average 11.3 cents per thousand cubic feet in 1957, 14 cents in 1960, 16 cents in 1967, 18.6 cents in 1972, the last full year before the OPEC oil embargo. In real (uninflated) dollars, the average price of gas from 1950 to 1972 rose only from 6.5 cents to 12.8 cents.

Yet controls, if they are to be equitable, must afford producers a reasonable return on their investment. Over the meaning of the term "reasonable," producers and regulators have squabbled throughout human history. To the regulators, a "reasonable" price is whatever seems socially acceptable, given the cost of production and their perception of the social utility of the product in question. Producers have a different view of the matter. Of course they wish their past costs reflected in the price, but also their future costs — the expense entailed in replacing, year after year, the product worn out or used up.

A fundamental trouble with controls is that they face backwards. They see what has happened before. But how can they predict the future, which is still to come? To changed conditions or new developments, controls are mulishly unresponsive, because their roots lie in the past. If demand for a product accelerates, causing a scramble for the equipment and the labor to expand production, the producer's costs climb in a way that no past statistics can account for. The most conscientious controller finds his data lagging months, if not years, behind the event.

At first, the Federal Power Commission's prices were adequate to encourage additional drilling. The year after the Phillips decision, 1955, gas drilling slumped by more than 300 wells — to 3,608. But the following year it recovered, and the upward drilling trend of so many years' duration continued on its way. There were 4,543 gas wells drilled in 1956; 4,620 in 1957; 5,029 in 1959. In 1962, with average prices at 15.5 cents per mcf, the industry sank 5,848 wells and discovered more than 19 trillion cubic feet of natural gas.

Yet 1962 proved the high water mark for gas-well drilling — at least until rising prices occasioned by the energy crisis reawakened interest in gas exploration. For 1963, the drilling count slumped by more than a thousand wells below the previous year; by 1967, it had dropped below 4,000. In 1968, a mere 3,456 gas wells were drilled, fewer than in any year since 1953 when demand was far smaller.

The trials and travail of one producer in North Texas were characteristic. Before a subcommittee of the House Interstate and Foreign Commerce Committee, D.K. Davis, at the time an independent producer associated with the Pitts Energy Group, testified that:

> Frank Pitts of Pitts Oil Co., which is one of the Pitts Energy Group companies, had two wells that sat about four or five miles to the west of our discovery well. His company had drilled those wells in 1967. They had never been hooked up . . . The producibility of those wells was so low that the pipeline company would not pick them up; would not build a pipeline to them. Consequently, they sat there for many, many years (in fact until 1973, when we signed a contract with Delhi Pipeline Co.) and eventually, those two wells were hooked up.
>
> In 1967 Mr. Pitts had stopped drilling. The price of gas was 17 cents. His bankers told him to quit, and he did. He would have liked to have had those wells hooked up, but the purchaser to which he was contracted would not put a pipeline into them.

To be sure, gas reserves for a time continued to rise. The wells being drilled, even if fewer in number, were mostly development wells which struck greater quantities of gas per well drilled. Though drilling began to drop off in 1963, the 20 trillion cubic feet discovered in 1964 exceeded the 19 trillion found in the peak drilling year of 1962 — and was as great as in the last year before the Phillips decision, 1953. Reserve discoveries peaked in 1967 at close to 22 trillion cubic feet.

Yet there was another, more unpleasant side to the coin. Fast as reserves might be added, they failed to keep pace with the growth of demand. The life expectancy of the nation's gas reserves grew shorter and shorter as demand grew higher and higher.

Whereas by year-end of 1950 only 184 trillion cubic feet of gas was in reserve, this was adequate for 26.9 years' use.

Ten years later, with reserves at 262 trillion cubic feet, the nation had enough gas to last only 20.1 years. The gap between supply and demand was narrowing fast. It was to narrow still faster during the '60s as demand, fed by low, federally dictated prices, rose sharply. The cheapness of natural gas throughout the '50s and '60s was a spur to homebuilders and plant owners looking for ways to cut energy costs. Early in the 20th century, crude oil had become the nation's principal industrial fuel, at least partly because abundant supplies rendered it cheaper in much of the country than coal. Now, very low prices made natural gas a popular commodity. Of course, from one perspective this was splendid: it meant high sales volumes for the natural gas industry and, for the nation as a whole, a more diversified supply of energy. By the principles of free market economics, increased demand would perpetuate supply; the "invisible hand" of Adam Smith would flex its fingers in the quest to find as much gas as the nation desired.

The trouble was that free market economics are dependent on the proper working of the price system. Prices are messages that travel back and forth between producers and consumers. Left alone by outside authorities, the two groups decide on the value of this commodity or that one. The producer's message is that it costs x-number of dollars to produce a certain commodity and that, for his time and investment, he expects y-per cent return on his investment. With which assessment the consumer is wholly free to agree or disagree. His purchase is a nod of agreement; his refusal to buy is a frown of disapproval. The marketplace — which is nothing but a name for the whole process of communications — thereby is said to set prices, and to set them at that point where the needs of consumers and producers intersect.

Supply is an additional ingredient in the price equation. The rarer a commodity, the more prized it becomes, and thus the more costly; conversely, the more abundant it is, the cheaper. Nothing is more plentiful than air, which, by no accident, costs nothing. As demand for a product rises, so does the producer's interest in meeting that demand. But — and a hugely important "but" it is — he must be free to test the market's willingness to accept the higher prices that finance the search for new supplies. Lacking, because of government regulation, the freedom to raise prices, the producer will settle for what supplies he can develop at the compensation afforded him. He will cut corners and hold down investment because government will not let him do otherwise.

Accordingly, it may be said that only under a free market do supply and demand ever rest in rough equilibrium. A fair and reasonable, as opposed to a government-dictated, price is what keeps them there.

This was precisely the difficulty with the natural gas market in the late '50s, and especially through the 1960s. It was government, not the interplay of supply and demand, that fixed prices. The backward-looking nature of the price-fixing process cannot account for constant changes in demand that affect supply. Like those 20th century French generals who were said always to be fighting the last war, the price controllers were ever responding — if they responded at all — to the *last* shift in demand.

And so by the beginning of the 1970s, the nation was beginning to experience a wrenching shortage of natural gas. In 1968, for the first time, production had exceeded the quantity of gas added to reserves. There was left in the ground no more than a 14.8 year supply of gas. Reserves took their final upward turn in 1970, after subsequent development drilling in the original 1968 discovery of Prudhoe Bay confirmed the existence of 26 trillion cubic feet of gas; this discovery amounted to almost 10 per cent of American gas reserves. Yet by itself, Prudhoe Bay could never restore the flush conditions of the '50s. Vast as it was, the field represented only a little more than a year's production for the United States. More to the point, economic and, in particular, environmental, problems have prevented to this day the building of a pipeline to carry the gas to the rest of the country. The gas wealth of Prudhoe Bay remains tightly locked up in the midst of an energy crisis.

Rather than relief, the '70s brought only more distress. Each year, beginning with 1970, U.S. natural gas reserves dropped by some 10 trillion cubic feet. They had touched the 200 trillion level by 1978 — a plunge of almost 50 per cent from the peak year of 1967. By not the slightest coincidence, the number of natural gas producers declined by about two thirds between 1956 and 1971.

Yet in the meantime, market forces were operating to alleviate, at least for part of the country, the constraints imposed by the growing shortage. The Phillips decision, dealing as it did with interstate commerce alone, excluded from price controls a major portion of the gas market — that gas sold within the states where it was produced and therefore classified as "intrastate" gas. Not even the Supreme Court submitted that Congress had meant to control the price of gas that never crossed state lines. Such gas sold at whatever price the market would bear, which, in the days of towering surpluses, was not great. Indeed, the irony is that intrastate gas, its price arising out of competition instead of the earnest meditations of government officials, at first sold cheaper than gas in the regulated interstate market. For example, in 1956 interstate gas, at an average 10.4 cents per mcf, was priced nearly twice as high as the intrastate variety, then selling for an average 5.8 cents. By 1965, when intrastate prices had climbed at last to the 10.4 cent plateau, interstate gas had ascended higher yet, to 15 cents.

Still, the price was ample to furnish the customers of intrastate gas — mainly Texans, Louisianans, and Oklahomans — with steady and dependable supplies. Intrastate reserves of 86 trillion cubic feet in 1963 (against 188 trillion in the interstate market) constituted a third of all the nation's proved gas supplies.

So it was also in 1967, the year that gas reserves peaked at 293 trillion cubic feet: two-thirds of all gas was interstate, a third intrastate. It was in fact the heyday of the interstate market, though none could know it at the time, particularly given all the distractions: the Six Days' War between Israel and the Arab nations; hippies; "escalation" in Vietnam; draft-card burnings. Nor was the drop in American gas reserves the following year so very visible — a mere 6 trillion cubic feet overall. The next year's plunge in reserves was more noticeable, being twice as large as that of the previous year. And by 1970 it was plain, or ought to have been, that reserves were sliding downward at a fearful rate — 10 trillion cubic feet that year (not counting Prudhoe Bay), 12 trillion the next, nearly 13 in 1972. A tumble of 16 trillion cubic feet took place in 1973.

Not that gas no longer was being found. It was. Yet something was happening to it that the price controllers would have been loath to acknowledge. Increasingly, the nation's new gas supplies were being sold to intrastate pipelines at the increasingly higher prices afforded by the free market. Back in 1968, 6.4 trillion cubic feet — 65 per cent — of newly discovered gas had been committed to interstate reserves. Suddenly, in 1970, the interstate share dropped like a stone to 3.5 trillion of 11.3 trillion cubic feet discovered (excluding Alaska). The plunge the following year was still sharper. Of 11.1 trillion cubic feet discovered, only 2.2 trillion cubic feet was dedicated to the interstate market.

A venerable piece of folk wisdom has it that one can lead a horse to water but not make him drink. No more could the Federal Power Commission make producers sell their gas to the interstate market if they were positioned to sell to the more lucrative intrastate market. Facts had to be faced: the federally fixed price was not attractive enough. Not when the growing scarcity of gas was leading the intrastate pipelines rapidly to bid up the price — to offer more than the interstate carriers were permitted to pay.

The price differential grew especially sharp with the coming of the energy crisis in all its bleakness and terror. In 1974, the average price paid domestic producers by interstate pipeline companies was less than 30 cents per mcf until, in December, it climbed to 32.6 cents. Contrast this with the price paid by intrastate carriers for new Texas gas. The lowest price paid all year was 37 cents in the third quarter. Indeed, the average price that quarter was 87 cents, the high $1.38. For the fourth quarter, the average new intrastate gas price in Texas rose to $1.17, with a high of $1.73. It was hardly surprising that, given the choice, producers should shun the controlled-price market, selling instead in the free market.

As time went on — and the energy crisis deepened — the temptation grew all the stronger. Average Texas intrastate gas prices were $1.32 per mcf during the first quarter of 1975; by year-end, they were $1.87; by the end of 1976, they had reached $1.99, with gas fetching on occasion as much as $2.33. Yet throughout 1975, the amount paid domestic producers selling across state lines averaged only 38.3 cents per mcf. By 1976, the intrastate price was tugging the interstate price higher and higher; mandates to set prices "in the public interest" might be all very well, but the Federal Power Commission was learning that "the public interest" no longer meant — assuming it ever did — the need to keep prices low. If the interstate market was not to starve for new gas, higher prices had to be forthcoming.

Small wonder that the intrastate share of American gas reserves (lower 48 states) grew progressively larger: from 31 per cent in 1963, to 33 per cent in 1970, 37 per cent in 1972, 41 per cent in 1974, and 47 per cent in 1976. If, in the period, 1967-77, intrastate reserves dropped 10.1 trillion cubic feet, the plunge in interstate supplies was more than ten times as great.

It was inevitable, given the roaring success of the free market in intrastate gas, that covetous eyes should be cast on it. In 1978, by Act of Congress, the distinction between interstate and intrastate gas was wiped out (though the governors of Texas, Oklahoma, and Louisiana challenged the law in federal court and in 1980 were still awaiting a decision). Henceforth the two species of gas — which always were the same in taste, smell, and quality, differing only in destination — would be treated as one. The essential point had nevertheless been made: price controls are folly; they do not help the consumer — in the long run, they make his life harder than it would have been without them. It is a fine thing to have cheap gas. But cheap prices without the gas — of what use are they to anyone?

APPENDIX TO CHAPTER 3 THE NATURAL GAS SHORTAGE ON GRAPHS

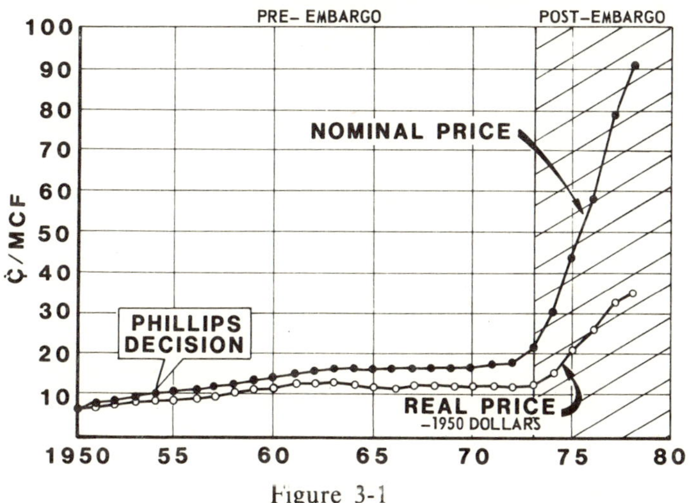

Figure 3-1

In 1950, the average wellhead price of U.S. natural gas was at the incredibly low level of 6.5¢ per MCF. By that time, natural gas had just begun to be a valuable resource in its own right, rather than the undesirable by-product of crude oil it had been in the pre-World War II era, when much of the gas not immediately used on location was being flared. By 1954, the average wellhead price of U.S. natural gas had risen to 10.1¢ per MCF. Following the Phillips decision of that year, the price of natural gas remained fairly stable until the year of the oil embargo (1973), when the price rose to 21.6¢ per MCF. By 1978, the price had risen to 91.9¢ per MCF. Also shown in Figure 3-1 are average natural gas wellhead prices after inflation adjustment, i.e., real prices in terms of constant 1950 dollars. The 1978 price, after inflation adjustment, was 35.9¢ per MCF.

Source: *Basic Petroleum Data Book,* API, Section VI, Table 2.

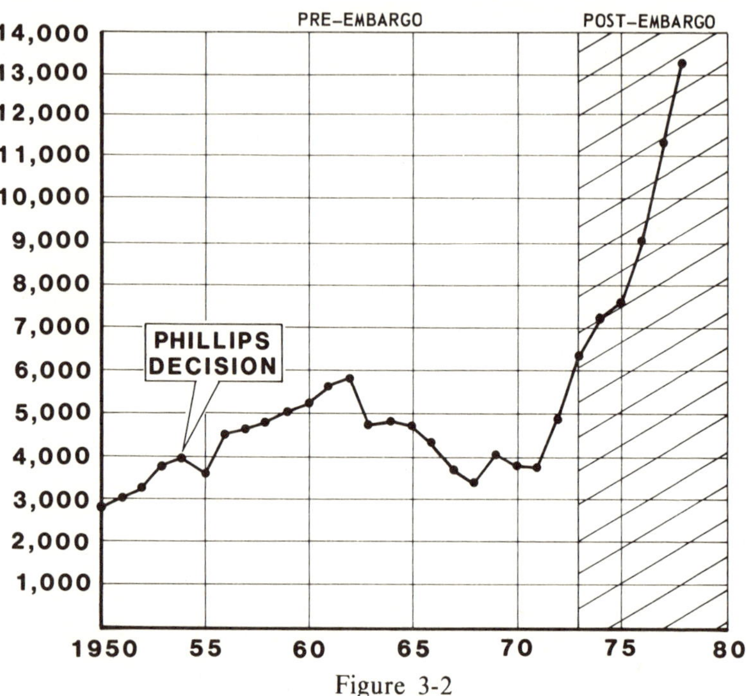

Figure 3-2

Rising from a total of 2,843 U.S. gas well completions in 1950, gas drilling activity had reached the all-time high of 3,974 wells in the year of the Phillips decision (1954). Ignoring the shock reaction in 1955, gas drilling activity continued to rise, reaching a high of 5,848 well completions in 1962. That was the second year in a six-year period when average wellhead prices were held at approximately 15¢/MCF. Result: gas drilling activities began to decline in 1963, dropping to 3,830 completions in 1971. Starting in 1972, i.e., well ahead of the oil embargo, natural gas well drilling activities began their drastic recovery until, by 1978, the total number of U.S. gas well completions had reached 13,064, more than double the previous high in 1962.

Source: *Basic Petroleum Data Book,* API, Section III, Table 2.

NATURAL GAS RESERVES AT YEAR-END

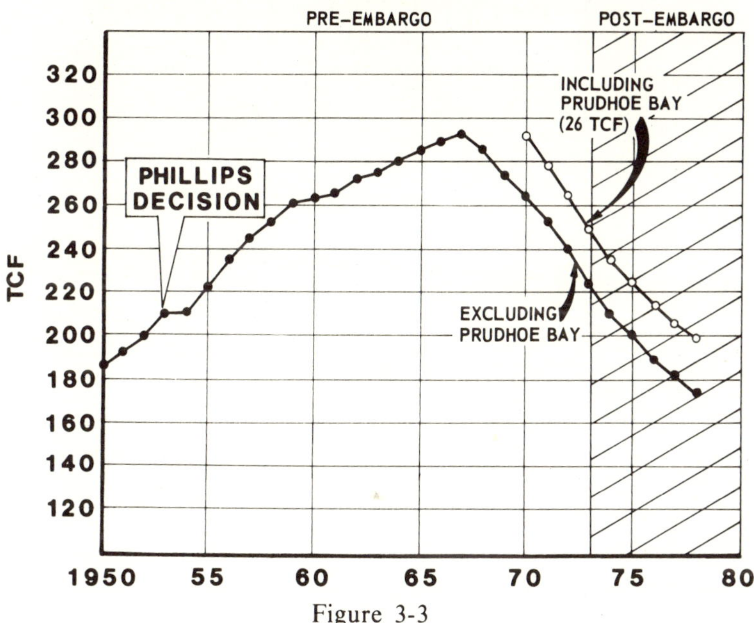

Figure 3-3

In 1950, proved U.S. natural gas reserves amounted to 185 trillion cubic feet (TCF). That's 185,000,000,000,000 cubic feet, enough to supply the United States for 27 years at 1950 consumption rates. Even though drilling activities began to decline in 1963, U.S. gas reserves continued to grow until 1967, primarily because the reduction in gas well drilling was at the expense of exploratory wells, while development wells in proven fields continued to be drilled with intensity. With low gas prices and unabated growth rates in consumption, the inevitable happened in 1967: gas reserves peaked at 293 TCF. By 1978, they had declined by 40% to 174 TCF, not counting the gas reserves in Prudhoe Bay, Alaska, that have not yet been made accessible to markets. Given the 1978 consumption rate of 19 TCF, the remaining reserves (including Alaska) represent no more than an 11-year supply. Note that the rate of decline of U.S. gas reserves is so drastic that a new giant Prudhoe Bay Field would have to be found every 18 months or so just to stay even.

Sources: *Basic Petroleum Data Book,* API, Section XIII, Table 2. *Reserves of Crude Oil, Natural Gas Liquids, and Natural Gas in the United States and Canada, as of Dec. 31, 1978,* API/AGA/CPA., Vol. 33, June 1979, Tables VII-1 & .2, pp. 116 & 117.

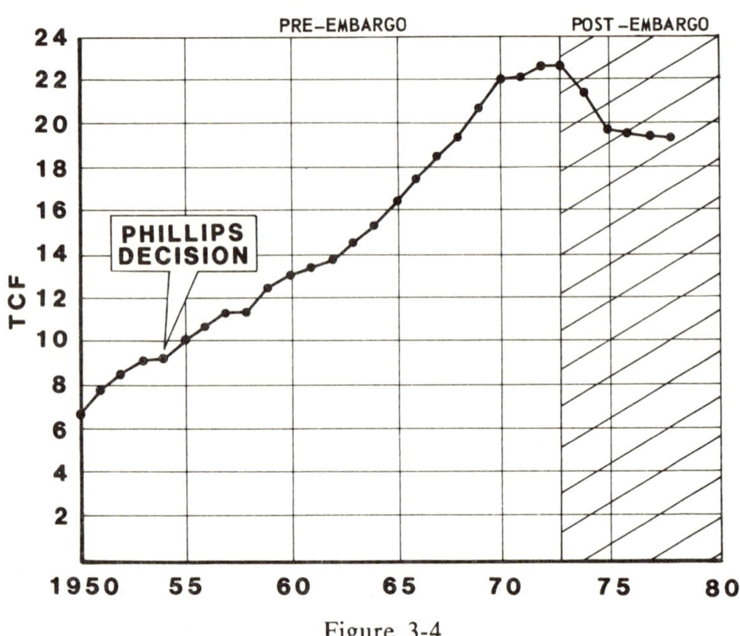

Figure 3-4

In 1950, U.S. natural gas production was 6.9 trillion cubic feet (TCF). With the wellhead price of natural gas controlled at ridiculously low levels, the demand for natural gas soared and U.S. gas production rose accordingly until, by 1973, an all-time high of 22.6 TCF had been reached, causing production to more than triple in 23 years. As was shown in Figure 3-3, unrelenting demand for natural gas led to net withdrawals from U.S. gas reserves. This temporary expedient of selling from inventories succeeded in keeping gas production rates at or above 22 TCF per year until 1973. By 1974, gas reserves had diminished to the point that they could no longer sustain these high rates, and production declined to 19.3 TCF by 1978.

Sources: *Basic Petroleum Data Book,* API, Section XIII, Table 2. *Reserves of Crude Oil,* etc., API/AGA/CPA, Vol. 33, June 1979, Tables VII-1 & .2, pp. 116 & 117.

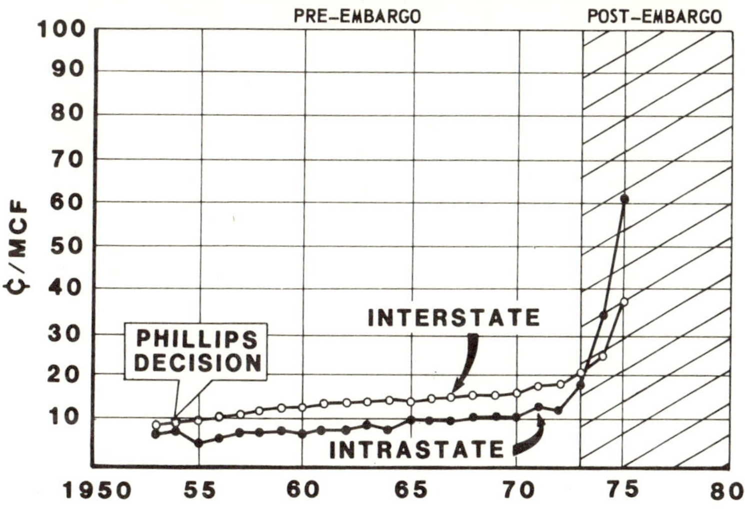

Figure 3-5

Since the Phillips decision affected only natural gas production committed to interstate commerce, the post-1954 period saw the development of two separate and distinct natural gas markets: the interstate gas market subject to federal price regulation, and the intrastate market covering natural gas that was produced and consumed in the same state. Prices in all intrastate markets remained free from regulation. As Figure 3-5 shows, average intrastate gas prices remained below interstate prices until 1973. When the natural gas shortage became critical in 1974, intrastate gas prices were free to respond. The magnitude of the intrastate price response is more dramatically demonstrated by focusing on a given year's new-contract prices rather than average prices, which contain old contracts at prices as low as 10¢ per MCF.

Source: *Texas Energy: A Twenty-Five Year History,* Governor's Energy Advisory Council, Report No. 77004, p. 102.

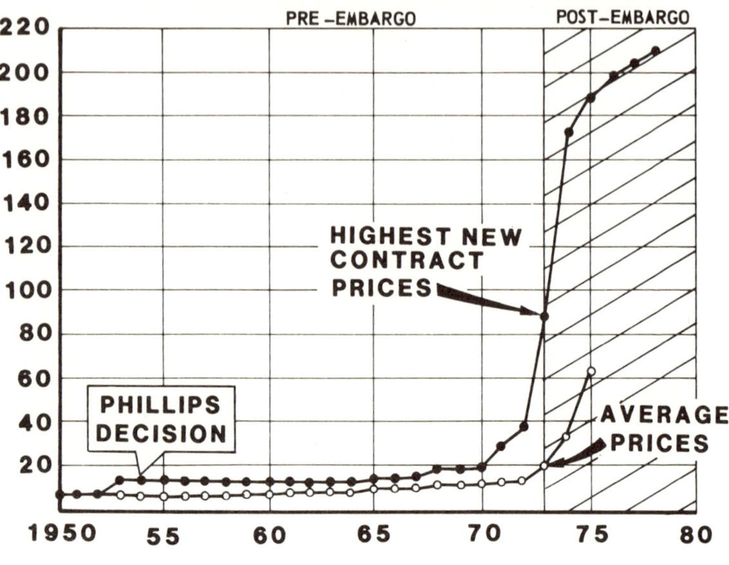

Figure 3-6

Actually, new-contract prices of intrastate natural gas picked up as early as 1971 when they rose to 30¢ per MCF, up from the previous year's price of 21¢ per MCF. By 1975, new-contract prices were as high as $1.90 per MCF. However, since new-contract commitments represented only a small portion of the intrastate gas market, they became negligible when averaged into the overall price. For example, the average intrastate gas price in Texas was 63¢ per MCF in 1975, compared to the previously mentioned new-contract price of $1.90 per MCF. While the contribution of new-contract prices was negligible in terms of average intrastate prices, it was certainly not negligible in terms of a motivating force in drilling new wells and developing additional gas reserves. Be this as it may, new-contract prices reached a high of $2.14 per MCF in 1978, the year in which the intrastate market was terminated by a legislative mandate extending price controls to all gas markets — including intrastate markets.

Sources: New-contract prices 1950-1975: private contacts. New-contract prices 1976-1978: *Monthly Energy Review,* U.S. Department of Energy.

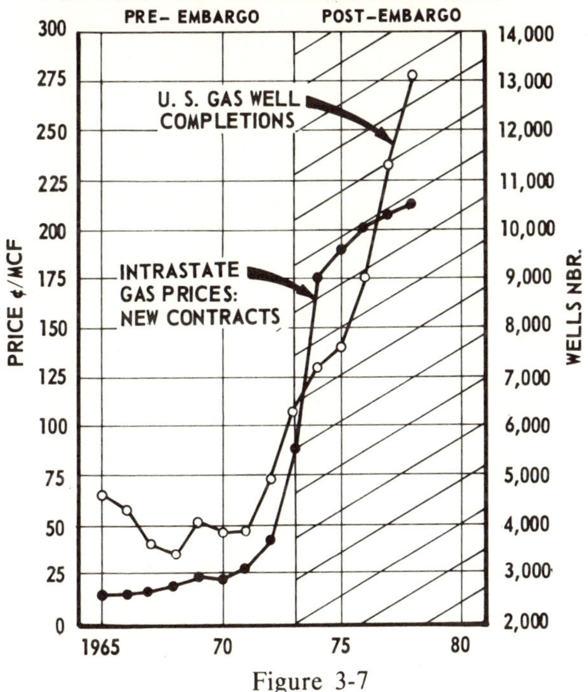

Figure 3-7

Producers do not drill gas wells on the basis of average prices. The decision to drill reflects the producers' estimates of future gas prices, and these estimates are heavily influenced by known prices of recently completed gas wells, i.e., by new-contract intrastate gas prices. While these prices remained at or below 21¢ per MCF until 1970, U.S. gas well drilling declined. 1971 was the year in which new-contract intrastate prices began their dramatic increase, and it was also the year in which gas well drilling budgets were revised throughout the industry. Beginning in 1971, the number of gas well completions jumped from 3,830 to 4,928 in 1972, to 6,385 in 1973, and so on. By 1978, a total of 13,064 gas wells were completed in the United States — more than double the number of the previous high (1962) under controlled prices. In all, the number of gas well completions more than tripled in the short span of seven years.

Sources: Figures 3-2 and 3-6.

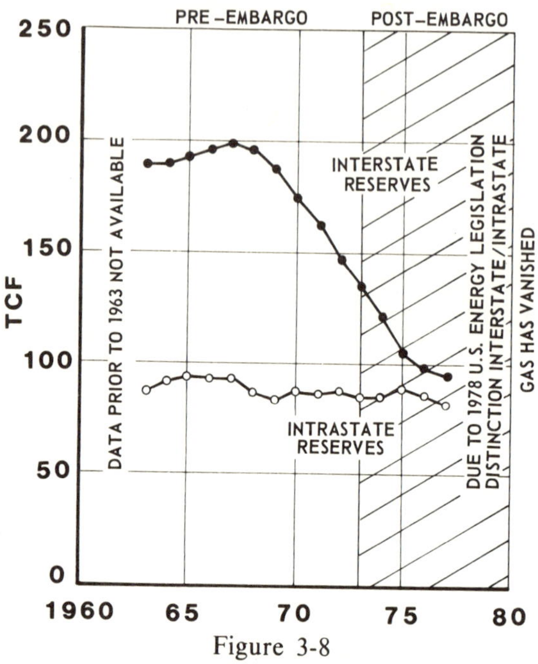

Figure 3-8

In 1963, the first year for which data are available on both interstate and intrastate gas reserves, total U.S. gas reserves amounted to 276 TCF. Of these, the interstate reserves were by far the most important. Estimated to be 188 TCF, interstate gas reserves represented 68% of the total U.S. reserves, or more than double the 1963 intrastate reserves of 86 TCF. When total U.S. gas reserves began to decline in 1968 (See Figure 3-3), this decline was due almost exclusively to reserve draw-downs in interstate gas markets. In contrast to intrastate gas reserves, which remained more or less the same over the period (1963-1977), interstate reserves peaked at 198 TCF in 1967, and subsequently declined to less than half as much 10 years later (1977). By that time, interstate reserves had almost come down to the level of intrastate reserves — not counting Alaska's proved but shut-in reserves of 26 TCF.

Source: *Gas Supply Review,* AGA Gas Supply Committee, November 1977, Vol. 6, No. 2, Table 1, p. 5.

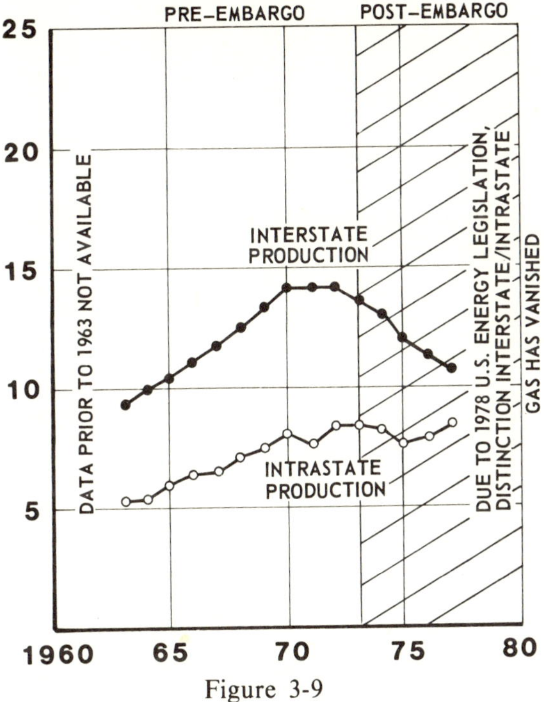

Figure 3-9

The rise in total U.S. gas production to 1973 was the result of increasing sales of both interstate and intrastate gas. As was pointed out in connection with Figure 3-3, however, the increase in overall gas production during 1967-1973 involved substantial reserve draw-downs. These draw-downs occurred exclusively among reserves committed to interstate markets. Beginning in 1973, interstate gas production could no longer be sustained by its shrinking reserve base until, by 1977, it had declined to 10.9 TCF. By contrast, intrastate gas production held its own at approximately 8.4 TCF per year, while the intrastate reserve base also stayed the same. Thus, the intrastate gas market was in equilibrium: at a price of approximately $2.00 per MCF, intrastate gas withdrawals were matched by equivalent volumes of reserve additions. Additional evidence of the effect of rising intrastate gas prices on drilling and reserve development is given in Figures 3-10 and 3-11.

Source: *Gas Supply Review,* APA Gas Supply Committee, November 1977, Vol. 6, No. 2, Table 1, p. 5.

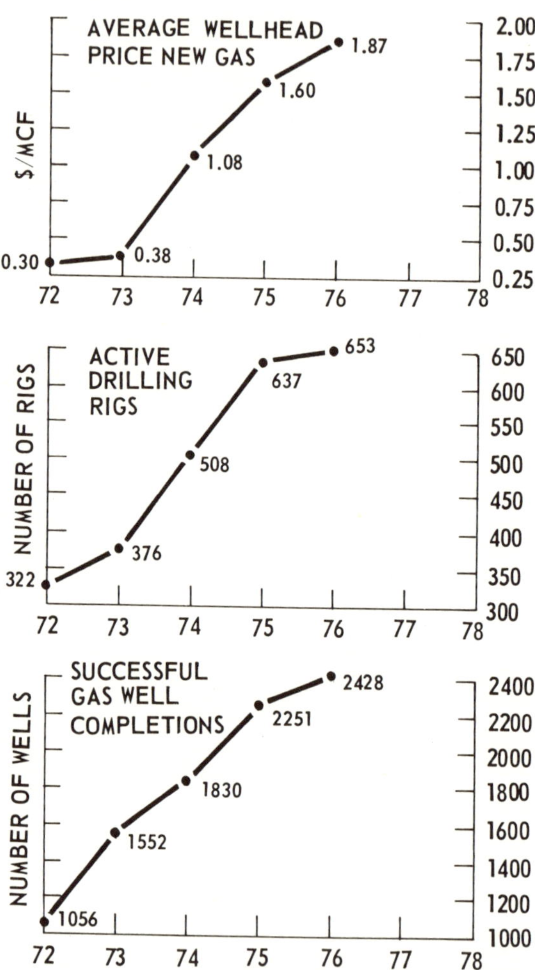

TEXAS : THE EFFECT OF FREE-MARKET PRICING ON DRILLING ACTIVITIES AND ON GAS WELL COMPLETIONS

Figure 3-10

Source: Resource Analysis & Management Group, Oklahoma City

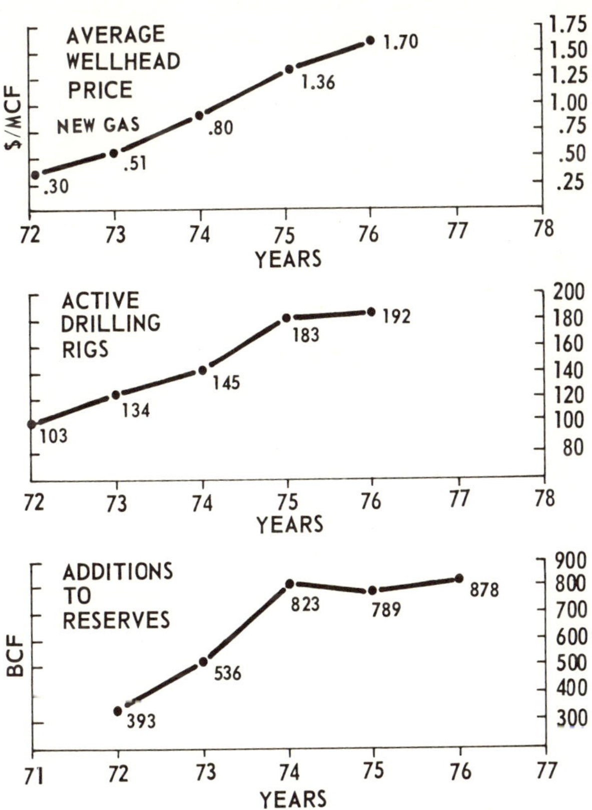

OKLAHOMA: THE EFFECT OF FREE-MARKET PRICING ON DRILLING ACTIVITIES AND ON GAS WELL COMPLETIONS

Figure 3-11

Source: Resource Analysis & Management Group, Oklahoma City

GROSS RESERVE ADDITIONS OF NATURAL GAS AND INTERSTATE RESERVE COMMITMENTS

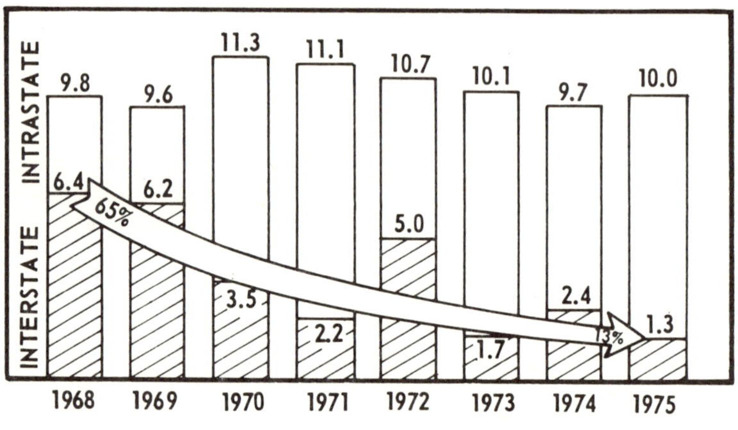

Figure 3-12

The oil and gas industry has been charged, (and possibly found guilty by the media and the public) with "withholding" low-priced interstate gas production. What really happened was that the producers' scarce drilling resources were increasingly devoted to the development of higher-priced intrastate gas reserves. For example, in 1968 and in 1969 approximately 65% of this nation's newly discovered gas reserves were committed to interstate markets. This percentage had declined to 13% by 1975 when new-contract intrastate gas prices had soared to $1.90 per MCF. Thus, the term "withholding" does not mean that interstate gas wells were systematically shut down. It means instead the non-replacement of slowly declining interstate gas reserves which were valued at prices below replacement cost.

Source: Sun Company

CHAPTER 4

THE CRUDE OIL SHORTAGE

The origins of the oil industry date back to a 69½-foot well completed on August 27, 1859, in Titusville, Pennsylvania, by a former railroad conductor, Edwin L. Drake. It was not such a well as would excite the slightest interest 120 years later. Its initial capacity was a mere 8 to 10 gallons a day. Yet it was the first of hundreds of thousands of wells to be drilled from Texas to Indonesia. Col. Drake, as the driller rather generously styled himself, had changed history.

Oil was an exciting commodity — liquid, easily stored, easily transported, and easily enough procured once technology advanced beyond Drake's rudimentary methods. Coal was the dominant fuel of the day, but oil had all the same its valuable uses — for lighting and even for medicinal purposes. (A testimonial letter, printed in a circular advertisement, paid tribute to petroleum's beneficial effects on "the most intense suffering and agony" caused by "Inflammatory Rheumatism.")

In due course, oil production spread from the seaboard states westward to Ohio and Kansas. The nation in 1900 produced 63.6 million barrels. A considerable industry was in place.

On January 10, 1901, on a small hill near Beaumont, Texas, there erupted the most dramatic gusher in history. The well was being drilled by Capt. Anthony Lucas, a mining engineer, on land where informed opinion held no oil would ever be found. A great column of oil, gauged ultimately at 75,000 barrels a day, exploded high into the air. Soon the hill, known as Spindletop, was covered with drilling rigs. A new age — the fuel oil age — had been born.

Oil was admirably suited for fuel. It burned far more cleanly than coal and produced more units of energy. So vast — and, accordingly, so cheap — was the supply of oil found in Texas that ships, trains and manufacturing plants began switching from coal to oil. Gasoline refined from oil ran many of the new motor cars being made. Coal was never wholly supplanted, of course, but for much of the world, including the United States, oil became the principal fuel.

As far back as 1908, there were predictions that oil would soon run out. Yet always, before it could do so, great new supplies were found — such as the mammoth East Texas oil field and the great reservoirs of the Middle East. Singularly blessed in so many natural resources, the United States proved to have a superabundance of oil. In the winning of two global conflicts, America's oil supplies were instrumental. After World War I, Lord Curzon, a member of Britain's war cabinet, asserted that the Allies had "floated to victory on a sea of oil."

Given the transcendent importance of the fuel, it remains hard, in retrospect, to see how the United States could have permitted its oil supplies to dwindle, its dependence on foreign sources to soar. Yet so it happened.

The essential problem at the outset was the price and availability of foreign crude oil. The costs of producing oil in such countries as Saudi Arabia and Venezuela was low — lower even than in the United States. When transportation costs were figured in, foreign oil still remained cheap. The oil companies in 1950 used imports to satisfy 13 per cent of the nation's demand for oil, and by 1970, the figure had grown to 23 per cent. With so much low-priced foreign oil to draw from, it was no surprise that domestic oil prices remained low. Oil that sold for $2.78 a barrel in 1954 scarcely rose above that level during the next 15 years. By way of contrast, consumer prices during the period rose by nearly a third.

That great, mysterious entity, the marketplace, is never slow to recognize a shift in the balance of risks and rewards. In 1955, the U.S. oil industry sank more oil wells than ever before in its history — 31,567. Then, in 1956, drilling began to slump. In 1957, only 28,000 oil wells were sunk, and the next year, 4,000 fewer than that. Throughout the early '60s, the annual drilling rate held steady at just over 20,000 wells; and then, in 1965, it began falling once more: down, down, until in 1973, the year of the oil embargo, only 9,900 oil wells were drilled in all the United States.

It was a regrettable time for domestic drilling to turn so sharply downward, because consumption, fed by the most sustained economic boom of the century, was headed in the other direction. From 1950 to 1973, the economy each year devoured 4.3 per cent more oil than in the previous year, so that by the time of the embargo, consumption was three times as high, 6.3 billion barrels, as at the start of the '50s.

Not that the oil industry, which is a business like any other, proposed to turn away its customers. These had to be satisfied. Existing reserves, which reached 31.7 billion barrels in 1959 before ceasing to grow, were pumped all the harder — and, not surprisingly, started to diminish. It was simple mathematics: where more is taken out than is put back in, the whole grows progressively smaller. In 1968, this "drawdown" of the nation's oil supply first became pronounced. There was a dramatic reversal in 1970 when the great Prudhoe Bay field in Alaska was brought in, adding some 10 billion barrels to the nation's oil supplies. But the following year, the drawdowns resumed and have gone on since then.

There was, of course, another way of satisfying customer demand: buying even more foreign oil than before. And this the oil companies proceeded to do. The more that foreign oil held American prices down, the more foreign oil it became necessary to buy, because the less American oil there was to meet the growth in consumption.

In 1973, the equation changed drastically; or at least the occasion for drastic change arose. In that year, the Organization of Petroleum Exporting Countries suddenly quadrupled the asking price of its oil. In a twinkling, the age of "cheap foreign oil" was gone the way of the Edsel and the 5-cent soft drink. The exporting nations had, up until then, failed to act in concert to influence prices; nor, so long as America enjoyed spare productive capacity, would it have done them any good to act in concert. The United States would simply have opened up its oil wells and forced the price back down. But by the early '70s, that spare productive capacity had disappeared. The United States, to say nothing of the rest of the world, was dependent on non-American oil in a way that was inconceivable in the 1950s. OPEC had the ability to ask what price it might; the world would pay up and try to look as happy about it as possible.

On the other hand, there was vast opportunity here. The enormous increase in foreign oil costs meant that domestic oil prices could at last rise, American oil now being underpriced in comparison with foreign; it meant that domestic drilling could turn upwards again, that America's increasing dependence on foreign oil could finally be broken. All these things *could* have come to pass. They did so only in a partial, unfinished sense. For which there was an imposing reason: in 1971, the philosophy of the Phillips natural gas decision — price control in the name of the public interest — had been extended to crude oil. It had been extended, in truth, to the entire American economy. On August 15, 1971, fearful of being attacked during an election year for failure to cope with inflation, President Nixon froze all wages and prices. Subsequently he permitted them to rise, but only slowly, and only in accordance with government-prescribed formulas. Happily for the rest of the economy, the controls experience was a short one. By early 1974, the controls program had been ended. But meantime there had occurred the oil embargo and the meteoric rise in foreign oil prices. Before the oil industry could be liberated of controls, the need to retain those controls on oil had become a part of political orthodoxy. The rest of the economy was freed — but not oil. In the nearly two decades since the Phillips decision, the government had forgotten nothing and learned nothing. It still believed that price controls operated for the public good, sparing the public the pain of sharp price rises, affording the industry at the same time enough profit and capital to go on about its business. That controls, applied to gas, had produced the gas shortage and would surely worsen the oil shortage was a point lost entirely on the controllers.

It was the first time in American history that oil prices had been held down by government action. On one occasion, they were held *up* by government action — when Texas, in the early 1930s, set production ceilings for the East Texas oil field. The frantic haste to pump the immense field dry was not only ruining the producers, who had to sell their oil at a few cents a barrel, but was depleting the reservoir pressure, by which the oil was lifted to the surface. At least in Texas, the emphasis was on greater returns for the producers; the emphasis in the early 1970s, owing to federal policy, was on comparatively modest prices for the consumer — never mind the consumer's compelling need for new supplies of oil to be developed as quickly as possible.

Once OPEC's members undertook to raise their prices, it was plain that something had to be done for the American oil industry or no more wells would be drilled. The government could not expect producers to settle for a price only a quarter of that permitted foreign sellers. And so a confusing formula — explicable only in political terms — was devised. Oil from leases being produced in 1972 was classified as "old" oil (the designation subsequently was changed to "lower tier") and priced at $5.03 a barrel. "New" oil (subsequently "upper tier") was that produced from discoveries since January 1, 1973. A price of $9.82 a barrel was allowed it in early 1974 — high by comparison with the "old" variety but well below the $11.59 a barrel that Arab oil commanded once the embargo ended and exports resumed. So American producers were to have an incentive all right, a spur to increased drilling activity and production. Yet what an incomplete incentive, what a dull spur it was!

The sinking of an oil well is an affirmation of belief and confidence. The driller believes that he may find oil; and he believes something more: that whatever oil he finds will yield him a satisfying profit on his investment, enabling him to drill yet another well and another and another. Faith, belief, confidence — these things are partly psychological, functions of the inner man. Yet they are also functions of graphs and charts and accounting books; which is to say that faith comes all the easier, and sticks all the more stubbornly, when the prospect of profit is there. For men to gamble their time, their money — perhaps even their lives — they must be sure of themselves and likewise sure of benefiting from all their work. Understanding the risks, they must believe the risks to be worth it.

That was precisely what American oil producers could not believe for certain in the mid-1970s. The element of risk is high in all business, but almost nowhere so high as in oil. After all, oil cannot be seen from above ground. Until the drill bit bites the earth, there can be no certainty that anything is down there at all. The success-to-failure ratio of oilwell drilling has traditionally been about one in ten. The ordinary exploratory well has, in other words, a 10 per cent chance of finding and producing oil. So what is wanted for the application of brains and capital to this risky business is the clear, unmistakable hope of a fair profit. If the prospect is not there — well, plenty of safer ways to invest money can be found. Taxfree municipal bonds, for example.

The term "fair" is one to be disputed between producers and price controllers. The controllers submit (historically, the breed has always submitted it) that the price they offer is fair both to producer and consumer. The assertion is not wholly without truth. For instance, the upper-tier price has in fact stimulated drilling. In 1974, for the first time in two decades, the drilling rate turned upwards. Nearly 13,000 oil wells were sunk that year, a gain of almost 3,000 over the previous year. As permissible prices rose, so did drilling. And in 1978, nearly 17,800 wells were drilled.

Which is progress to be sure, but an additional question is raised: absent controls of any sort, how much higher would the drilling rate rise, how much more oil would be added to the nation's reserves? The probable answer to both questions is: a lot.

Some hard-and-fast indications may be gleaned from considering stripper oil — the ten or less barrels a day that nearly depleted wells are able to yield. Under the Energy Conservation and Production Act of 1976, stripper wells were exempted from price controls, which were deemed too heavy a burden on such marginal operations. By April, 1979, the price of stripper oil had risen to $14.88 a barrel, virtually the same as OPEC oil when landed in the United States. It seems strange to reflect that a free market in petroleum exists only for tiny wells near the point of expiration.

At all events, as the Interstate Oil Compact Commission and the National Stripper Well Association reported late in 1979, "the free market price being extended to these marginal wells is extending their economic life." It would be vain to look for sizable increases in production from individual stripper wells; their capacity is minimal to begin with. The relevant figure is *abandonments.* Whereas in the period 1969 to 1975, under price controls, between 13,000 and 18,000 strippers were abandoned each year as unprofitable; only 8,380 were shut down in 1978, after price controls on stripper oil had been removed. The higher prices had given producers an incentive to go on pumping oil that otherwise would have been left below ground. Nor is the quantity of oil involved so unimportant as might be supposed. The nation's stripper wells, in 1978, produced 391 million barrels of oil, which, as the Interstate Oil Compact Commission's executive director noted, was 391 million barrels the United States had no need to import. Had controls *not* been removed, thousands of these wells — producing an average 2.86 barrels a day — would have ceased to function.

No such demonstrations of success are possible elsewhere in the petroleum industry, given that no other segments of the industry have been totally deregulated. Yet all studies with which the authors are familiar point clearly to the beneficial effects of price decontrol.

Writing in the magazine *Regulation,* published by the American Enterprise Institute, Nobel economics laureate Kenneth J. Arrow of Stanford University and Joseph P. Kalt of Harvard University note that the cost of controls on crude oil prices may be as high as $4 billion a year, given the uncertainties that they engender in the producers' minds.

Melvin R. Laird, former congressman, secretary of defense and domestic counsellor to the President, writing in *Energy — A Crisis in Public Policy*, American Enterprise Institute, notes that "continued price regulations . . . have dampened incentives to find and produce more domestic oil . . ."

And William E. Simon, first director of the Federal Energy Office, former Secretary of the Treasury and author of the best-seller *A Time for Truth*, writes in *Policy Review* magazine, ". . . the result of holding down oil profits while prices are rising would be to cut the automatic link between higher prices and increases in supply. New oil would be less profitable and so less likely to come into the market."

Prior to the announcement of President Carter's phased decontrol program, Milton R. Copulos, energy analyst for the Heritage Foundation in Washington, D.C., argued that, "Depending on the form, lifting the lid on oil prices could result in reductions of import requirements of between 160,000 barrels per day and 800,000 barrels per day this year, and as much as 1.7 million barrels per day by 1985."

The oil companies, naturally, have projections of their own. Union Oil's 1979 estimate pictured an extra 1,300,000 barrels a day by 1985 in the event lower-tier oil were decontrolled in phases. Conoco, using a more complex pricing formula, projected a slightly lower 1985 addition — 1,136,000 barrels a day.

On one point the authorities are united: higher prices and higher production levels go hand in hand. The question is not whether; it is how much. By the logic of the free market, this should have been plain long ago. Indeed, one suspects it was always a great deal plainer than many decontrol opponents made out. The reason that Americans have handed over whatever amount OPEC might demand, while denying a comparable sum to American oil producers, has in part to do with necessity: Americans, unlike Arabs, must obey the mandates of Congress and the President. In part, the reason has to do with resentment, dislike, even outright animosity: The oil companies, because they sell oil, are blamed for the high price of their product — as if, after more than a century of supplying low-cost energy, they had maliciously turned on their customers, fearless of the ugly consequences, which include congressional wrath, tighter government controls, bills to break up the integrated companies and populate their boards with "consumer spokesmen." To have blamed the oil companies at the start of the energy crisis may have been an understandable — however regrettable — instinct. To go on blaming them year after year makes no sense. Of course, sense has not been the common denominator of American political responses to the energy crisis.

APPENDIX TO CHAPTER 4 THE CRUDE OIL SHORTAGE ON GRAPHS

AVERAGE WELLHEAD PRICE OF U.S. CRUDE OIL

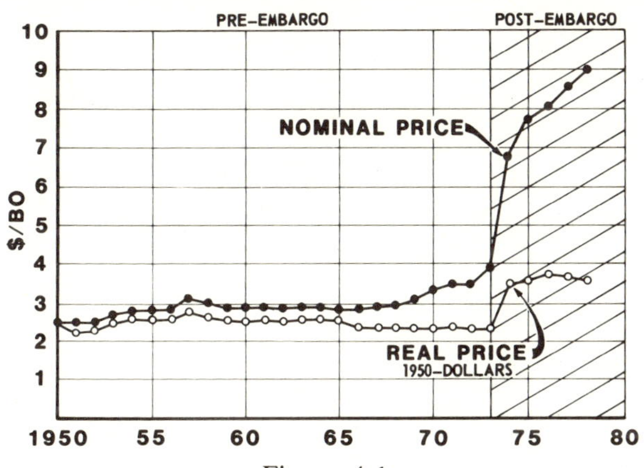

Figure 4-1

In 1950, the average wellhead price of U.S. crude oil was $2.51 per barrel. By 1954, crude oil was selling for $2.78 per barrel, and that price remained more or less the same until, by 1969, it rose past the three-dollar mark, never again to drop below it. The oil-embargo year of 1973 found the average wellhead price of U.S. crude oil at $3.89 per barrel. By 1974 it had risen to $6.74, and by 1978 it was $9.00 per barrel. It was the availability and low price of foreign crude-oil, mostly developed, produced and marketed by U.S. oil companies, that kept the domestic price of crude oil in check. The removal of this price constraint by OPEC induced a drastic increase in domestic crude-oil prices in 1973. Adjustment for inflation on the basis of the producer price index shows that the real price of crude oil (in constant 1950-dollars) declined slightly from $2.51/BO in 1950 to $2.36/BO in 1973. The constant-dollar 1974-price increase turns out to be only a little more than one-dollar; and after 1976, the real price of U.S. crude oil began to decline again in the midst of a severe energy shortage!

Sources: *Basic Petroleum Data Book,* API, Section VI, Table 1, and *Monthly Energy Review,* U.S. Dept. of Energy.

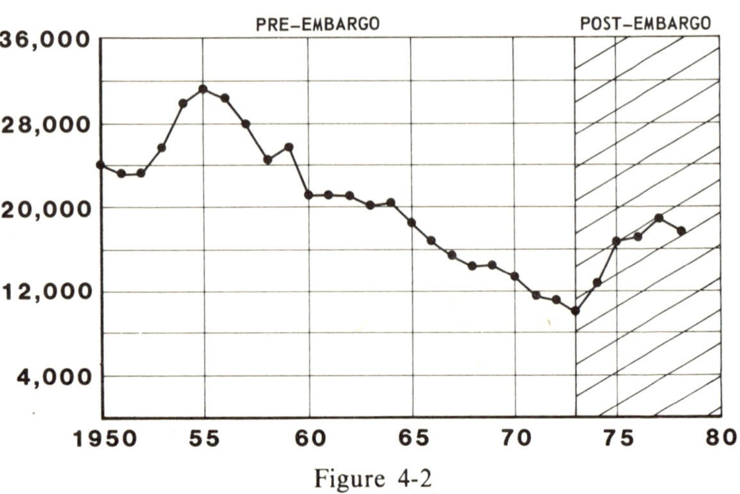

Figure 4-2

Starting from 24,430 oil well completions in 1950, drilling reached an all-time high of 31,567 oil wells in 1955. Thereafter, U.S. oil could no longer compete with inexpensive foreign oil, and drilling declined for nearly 20 years until, in 1973, fewer than 10,000 oil wells were completed in the United States. In contrast to natural gas, where direct foreign competition was for all practical purposes nonexistent, and where a price response from the intrastate markets caused drilling efforts to pick up at least two years before the embargo, the number of oil wells drilled in the United States did not begin to rise until after the embargo. By 1974, oil well completions had risen to 12,780 from the preceding year's 9,900. By 1978, nearly 17,800 oil wells were completed in the United States. That's still only about two thirds of the 1955-high, suggesting that drilling needs to be stimulated further.

Source: *Basic Petroleum Data Book,* API, Section III, Table 2.

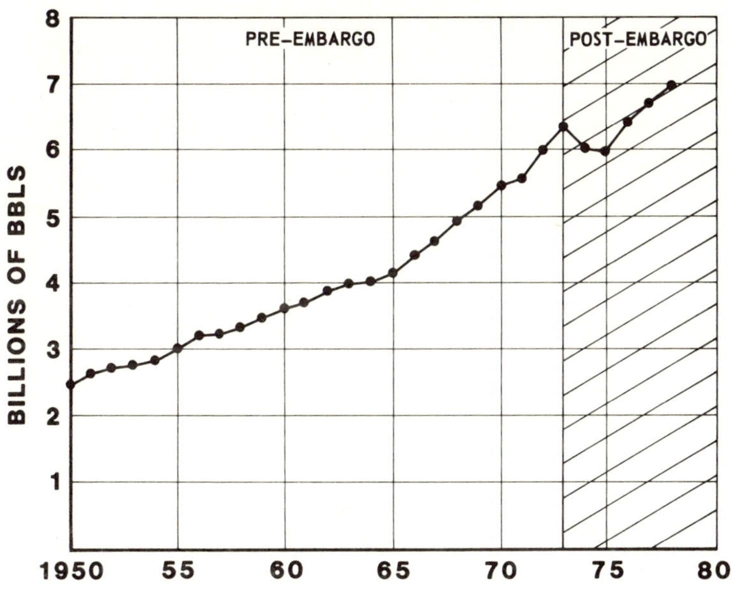

Figure 4-3

U.S. oil consumption, including lease condensate and natural gas liquids in addition to crude oil, has shown a remarkably steady growth pattern. From 1950 to the 1973 embargo, consumption grew at the rate of 4.3% annually. Starting at 2.4 billion barrels in 1950, consumption nearly tripled to 6.3 billion barrels in 1973. In 1974, U.S. oil consumption declined to 6.1 billion barrels; another small decline in oil consumption followed in 1975, and thereafter the shock impact of the embargo wore off. Abetted by domestic price controls on crude oil, consumption re-assumed its growth pattern with a vengeance: compared to its earlier growth rate of 4.3% per year, the post-1975 period exhibits an annual growth rate of 5.3%, despite exhortations from the U.S. government concerning the need to conserve energy.

Note: Includes Lease Condensate and Natural Gas Liquids.

Source: *Basic Petroleum Data Book,* API, Section VII, Table 3.

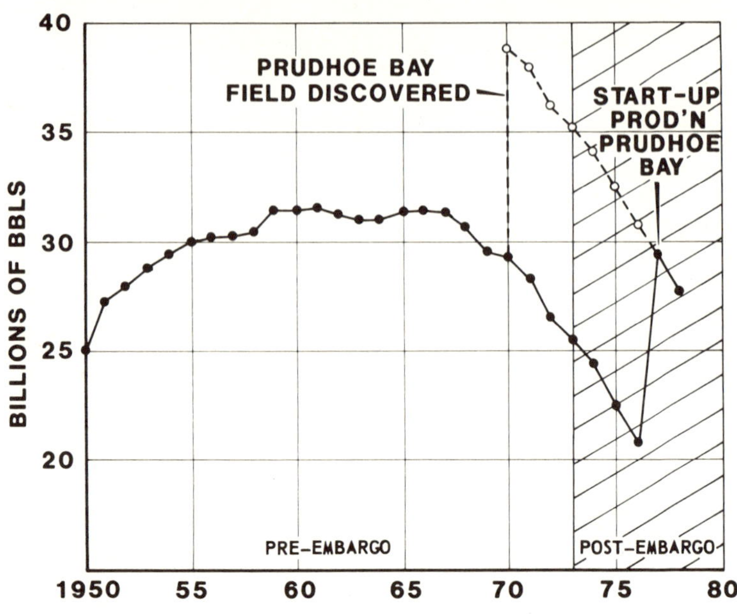

Figure 4-4

In 1950, U.S. crude-oil reserves amounted to 25.3 billion barrels at year-end. At 1950-production rates of 1.9 billion barrels per year (not counting natural gas liquids), U.S. crude-oil reserves were the equivalent of 13 years' worth of production at that time. Crude-oil reserves rose until 1959, when they reached a high of 31.7 billion barrels. Since crude-oil production had risen more or less proportionately to its reserve base, the 1959 U.S. crude-oil reserves still represented approximately 13 years' worth of production. During the period 1959 to 1967, U.S. crude-oil reserves held their own at levels slightly above 31 billion barrels. However, since production continued to rise during that period, the reserve base began to shrink relative to the production rate it had to support until, by 1968, U.S. crude-oil reserves corresponded to less than 10 years' worth of production. The reserve decline after 1968 was inevitable and dramatic: in spite of the one-time addition of the eighth-largest oil field ever discovered anywhere in the world (Prudhoe Bay), U.S. crude-oil reserves *declined* to 27.8 billion barrels in 1978.

Source: *Basic Petroleum Data Book,* API, Section II, Table 2.

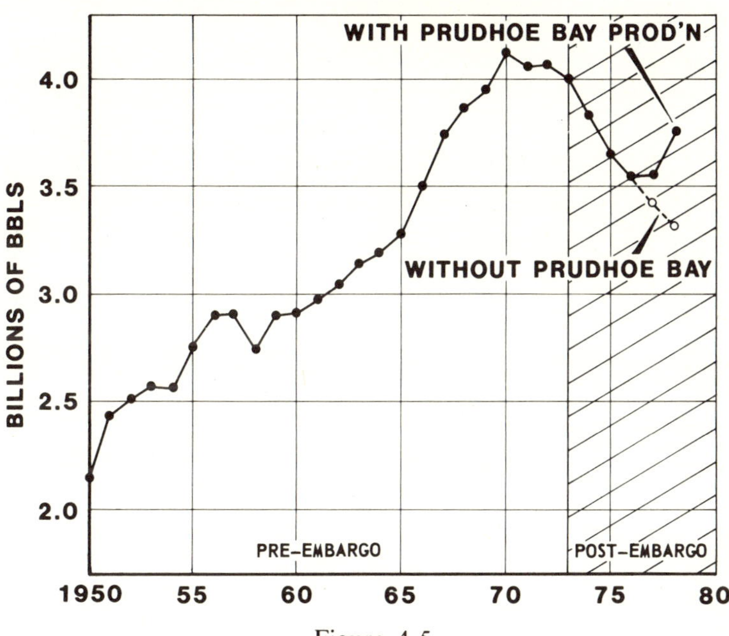

Figure 4-5

In 1950, U.S. crude-oil production, including lease condensate and natural gas liquids, amounted to 2.2 billion barrels. Even though drilling activities peaked in 1955 and the U.S. crude-oil reserve base stabilized four years later, U.S. oil production continued to rise, reaching 4.1 billion barrels in 1970. Clearly, when exploratory efforts decline while production rises, something has to give: not unlike the interstate gas situation, crude-oil reserves started to decline. In an attempt to provide for current needs, the U.S. began to undercut its capacity to provide for future needs. By 1971, the U.S. oil reserve base had shrunk to the extent that production rates could no longer be maintained. By 1976, oil production had declined to 3.6 billion barrels when some relief was provided through the Alyeska Pipeline: in 1978, oil production rose to 3.8 billion barrels. However, this was a one-time event. Barring the discovery of unforeseen major oil reserves within marketable distances, U.S. crude-oil production will resume its annual decline of 200 million barrels.

Note: Includes lease condensate and natural gas liquids.

Source: *Basic Petroleum Data Book,* API, Section VII, Table 3.

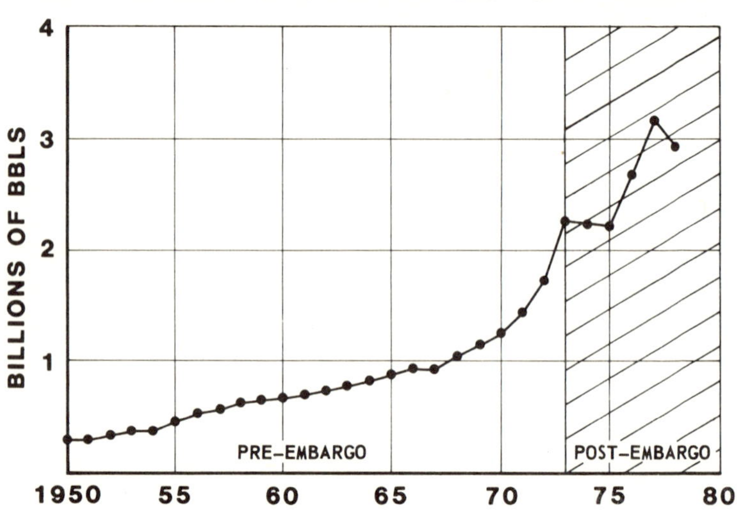

Figure 4-6

Unlike natural gas, crude oil is easily transported by sea. As a result, oil imports have played a crucial role in the overall U.S. energy picture. Starting at the modest level of 0.3 billion barrels in 1950, U.S. oil imports grew almost imperceptibly until, by 1968, they had reached the one-billion-barrel mark. Thereafter, coinciding with persistent U.S. reserve draw-downs, U.S. imports accelerated, reaching 2.3 billion barrels in 1973 — the year of the oil embargo. A minuscule two-year decline in crude-oil imports in 1974 and 1975 was largely due to reduced consumption. In 1976 crude-oil imports rose by nearly half a billion barrels, and in 1977 they again rose by a similar amount. In 1978, Alaska's newly-opened Prudhoe Bay field contributed some 450 million barrels of crude oil to the U.S. energy supply, causing oil imports to come down by approximately 230 million barrels. This temporary decline in imports was, for all practical purposes, arrested in 1979.

Note: Includes refined products

Source: *Petroleum Data Book,* API, Section IX, Table 1.

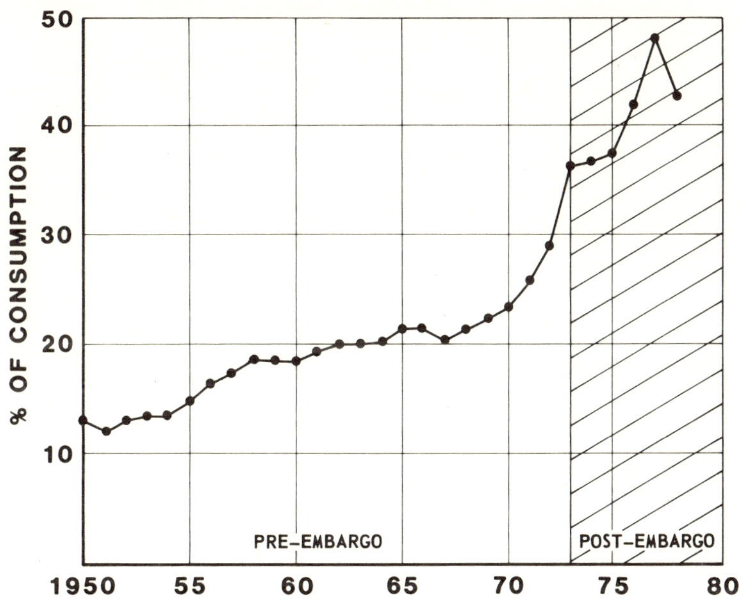

Figure 4-7

As mentioned earlier, when consumption exceeds production, the resulting shortage can be covered in two ways: by importing and/or by selling from inventories. U.S. oil imports, the visible part of the crude-oil shortage, amounted to 13% of U.S. oil consumption in 1950. Twenty years later, imports corresponded to 23%. Three years after that, it was 36%. Since U.S. oil consumption in 1974 and 1975 declined by more than imports, U.S. oil imports as a percentage of oil consumption continued to rise right through the post-embargo years, reaching nearly 48% by 1977. The opening of the Alyeska Pipeline and its impact on oil imports have already been discussed. Suffice it here to point out again that this one-time event cannot by itself reverse the current unfavorable trend.

Source: *Petroleum Data Book,* API, Section IX, Table 1.

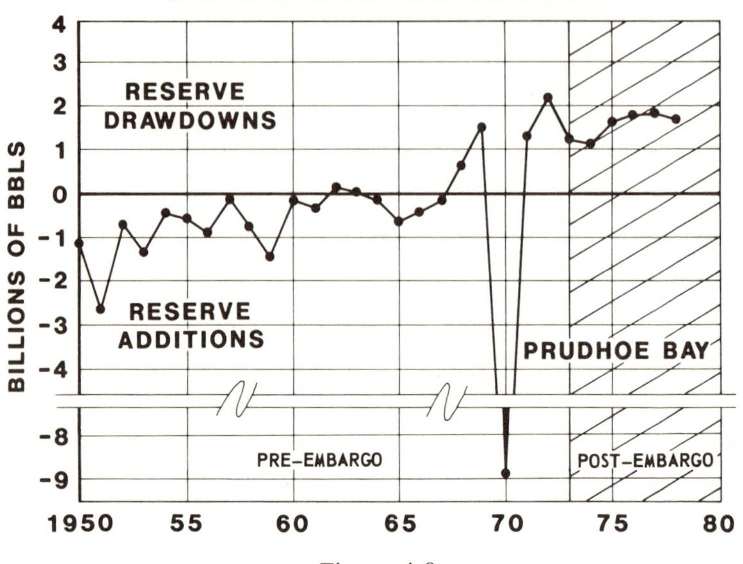

Figure 4-8

The predominant method used to cover the U.S. natural gas shortage was to draw down gas reserves, while imports played a relatively minor role. In crude oil, these roles are reversed: imports were the early and the important policy tool, while reserve draw-downs (the hidden shortages) came later and were smaller in magnitude. Taking into consideration both crude oil and natural gas liquids, two minor reserve draw-downs occurred in 1962 and 1963. It was not until 1968, however, that significant and successive draw-downs made their appearance: 0.7 billion barrels in 1968, 1.5 billion barrels in 1969, and somewhere between 1.1 and 2.2 billion barrels every year thereafter (with the exception, of course, of 1970, when the giant Prudhoe Bay Field provided a net addition of nearly 9.0 billion barrels).

Note: Includes natural gas liquid reserves

Sources: *Basic Petroleum Data Book,* API, Section II, Table 2, and *Reserves of Crude Oil,* etc., API/AGA/CPA, Vol. 33, June 1979, pp. 120 & 121.

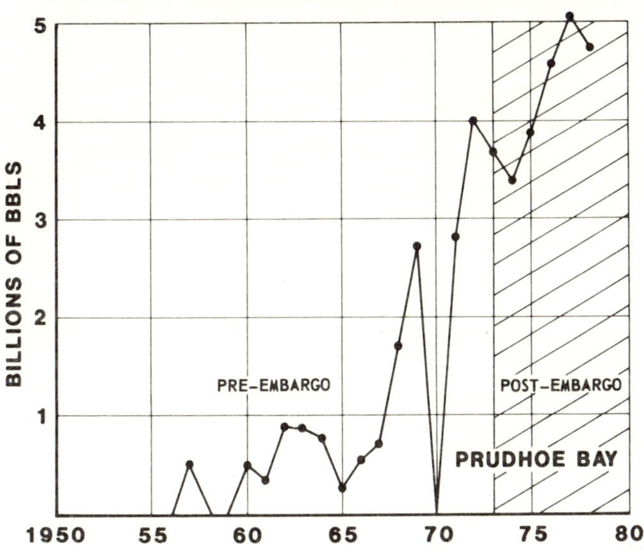

Figure 4-9

The total oil shortage in the United States covers both imports and reserve draw-downs, i.e., the visible and the hidden shortages. Apart from the one-time exception in 1957, there was no oil shortage in the United Stater prior to 1960, since relatively minor import volumes were more than compensated by reserve build-ups. One way of putting it is that the U.S. oil industry was preserving its domestic reserves at the expense of foreign reserves, and that is precisely the way it was viewed by the founders of OPEC. Be this as it may, once the oil shortage made its appearance, it grew seemingly without bounds. Apart from the Prudhoe Bay year (the field was discovered in 1968, but not clearly delineated until 1970), U.S. oil shortages were part of the American scene year after year. A modest shortage of one-half billion barrels in 1960 had grown to 4 billion barrels 12 years later. The immediate embargo years of 1973 and 1974 showed some minor relief, but that was temporary. By 1977, the total oil shortage reached an all-time high of 5.1 billion barrels. (continued next figure)

Source: *Basic Petroleum Data Book,* API, Section IX, Table 1 & Section II, Table 2.

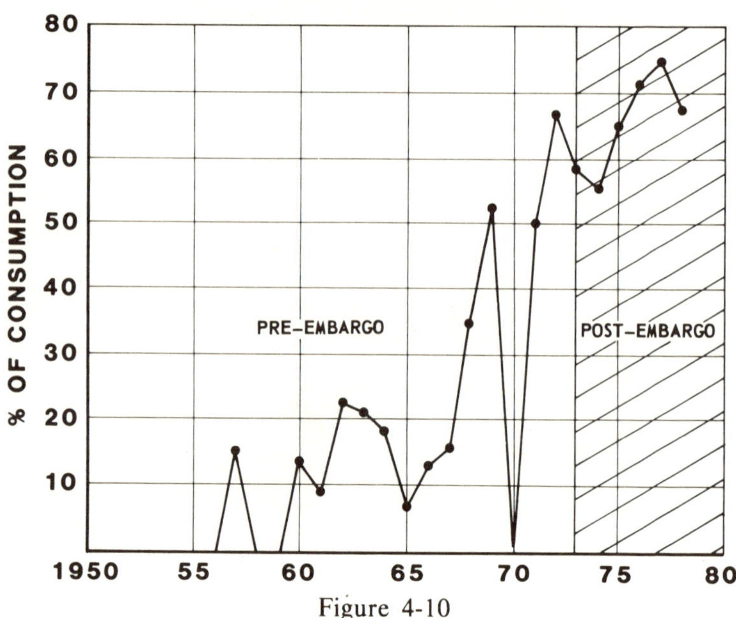

Figure 4-10

Notwithstanding two major warnings (the 1973/74 embargo and the short-lived but severe gas shortage in the winter of 1977), the U.S. energy shortage was permitted to continue its cancerous growth. From a zero oil shortage in 1954, the United States had developed a 5.1 billion barrel shortage by 1977. That is the equivalent of 14 million barrels per day! The Alyeska Pipeline, opened to full capacity in 1978, reduced the U.S. shortage by less than half a billion barrels a year. Since Alyeska is now operating at full capacity while a second pipeline is not in sight, this will be a one-time relief — hardly capable of reversing a generally and rapidly deteriorating trend.

More sobering than the oil shortage in terms of barrels is the oil shortage in terms of percent of domestic consumption. Starting at zero in 1959, the oil shortage had zoomed to 66% of domestic consumption 13 years later. Even during the embargo, the shortage remained above 50%, only to climb to 75% in 1977. The Alyeska Pipeline reduced the shortage to 68% of domestic consumption, but for how long?

Source: *Basic Petroleum Data Book,* API, Section IX, Table 1 & Section II, Table 2.

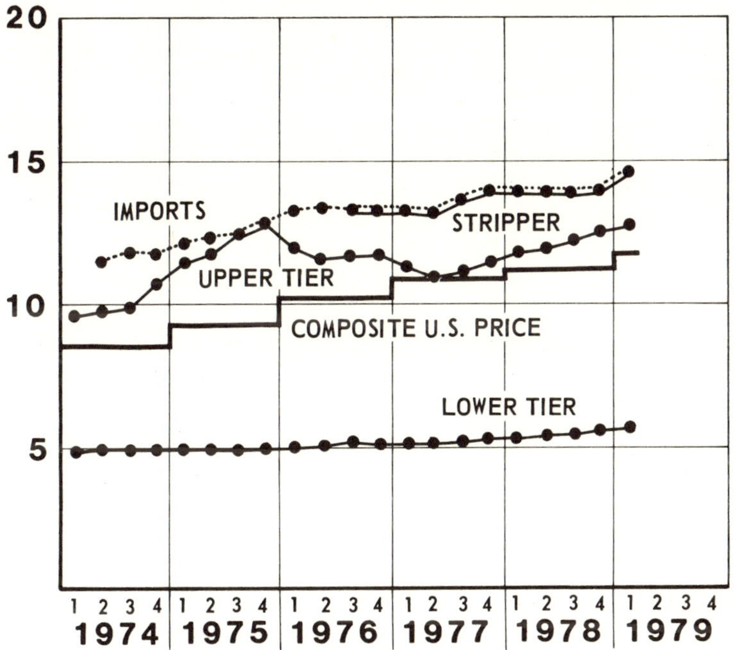

Figure 4-11

Post-embargo crude-oil price regulation in the United States has been confusing and futile, at least in terms of the stated objective of reducing U.S. dependence on foreign sources of energy. Several of the more important U.S. crude-oil classifications, each with its own set of price regulations, are shown above. Lower tier oil (previously called old oil and covering oil from leases that were on production as early as 1972) was originally priced at $5.03 per barrel, U.S. average; in April of 1979, the wellhead price of lower tier oil had risen to $5.85. *Upper tier oil* (or new oil under the previous terminology) was priced at $9.82 per barrel in early 1974. This oil reached a high of $12.99 in January of 1976, was subsequently rolled back to $11.39, i.e., by $1.60 per barrel and, at $12.93, had not yet reached its previous high by April, 1979. *Stripper oil* is oil coming from low-productivity wells, i.e., wells that are generally near abandonment. To preserve the life of these wells, stripper oil was deregulated under the Energy Conservation and Production Act of 1976. The price of stripper oil has been very close to the U.S.

Figure 4-11 (cont.)

landed cost of imported crude oil; as of April, 1979, stripper oil prices averaged $14.88 in the United States.

The *imported oil* shown here reflects the U.S. Dept. of Energy's estimate of the average landed cost of imported crude oil from Saudi Arabia. Starting with a price of $11.59 per barrel in April of 1974 (the first available data point), the price of imported oil rose to $15.28 in March of 1979. Only Mexico and Venezuela delivered crude oil to the United States at slightly lower prices at that time ($14.81 and $14.60, respectively), while most oil-exporting countries sold their petroleum at higher prices: Algeria at $16.61; the United Arab Emirates at $17.10; Canada (!) at $17.81; Libya at $17.56; and Iran at $19.54, to name some. Even the United Kingdom ($15.91) was selling its oil to the U.S. at prices higher than those charged by Saudi Arabia. Not included here are foreign spot prices that have reputedly gone as high as $50 per barrel.

Also shown in Figure 4-11 is the composite U.S. price of crude oil. This is an estimate of the weighted average of U.S. wellhead prices and foreign landed costs. Starting at $8.65 per barrel, the composite U.S. price rose to $11.71 in the first quarter of 1979. To be noted in connection with the composite U.S. price is that it has been moving away from the lower tier price. This means that the lower tier oil is failing to provide the intended price "protection." This is so because lower tier oil is declining and, as a result, becoming less and less important in the U.S. crude-oil sector. Originally estimated at 60% of the total U.S. crude-oil supply (64% in November of 1975!), it had declined to 34% in early 1979. Sooner or later, all of the lower-tier oil will be gone, and when that happens, the U.S. consumer will be more directly exposed to foreign crude-oil prices than he is now. Meanwhile, though, with lower-tier oil gone, an excellent one-time opportunity to develop alternative U.S. oil resources, conventional or not, will be gone as well: the capital that could have been generated for that purpose from the lower-tier/upper-tier price differential. In 1978, the funds that could have been (but have not been) generated this way were somewhere around $8.0 billion. Of that amount, roughly $4.0 billion would have accrued to the Federal Government through ordinary corporate income taxes; this could have been used for insulation subsidies, etc.

Figure 4-11 (cont.)

The other $4.0 billion, left in industry hands, would have drilled some 16,000 additional wells.

Sources: *Monthly Energy Review,* U.S. Dept. of Energy.

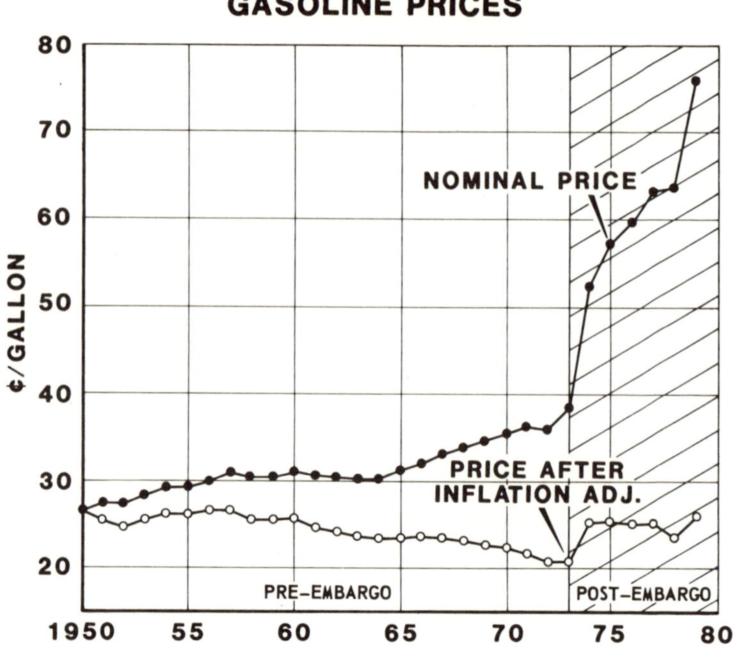

Figure 4-12

In 1950, the average U.S. price of regular leaded gasoline, with tax, was 26.8¢/gallon. Rising slowly over the years, it was still less than 40¢/gallon in 1973, the year of the oil embargo. With foreign crude-oil prices rising steeply and rapidly, while the production of lower-priced domestic crude oil was on the decline, the nominal price of U.S. gasoline rose to 52.4¢/gallon in 1974, to 57.2¢/gallon in 1975, and so on, until it reached 76.6¢/gallon in April of 1979. That is still less than half the price of gasoline in most European countries that are almost totally dependent on oil imports. Taking into account the general U.S. inflation on all prices, the price of gasoline after inflation adjustment (Consumer Price Index) declined over the 23-year pre-embargo period (1950-1973) from 26.8¢/gallon to 21.0¢/gallon. In March of 1979, the *real* price of gasoline in the U.S. (26.1¢/gallon) was still less than what it had been 29 years earlier!

Sources: 1950-1977: *Basic Petroleum Data Book,* API, Section VI, Table 4. 1978 & 1979 (April). *Monthly Energy Review, DOE,* July 1979, p. 83.

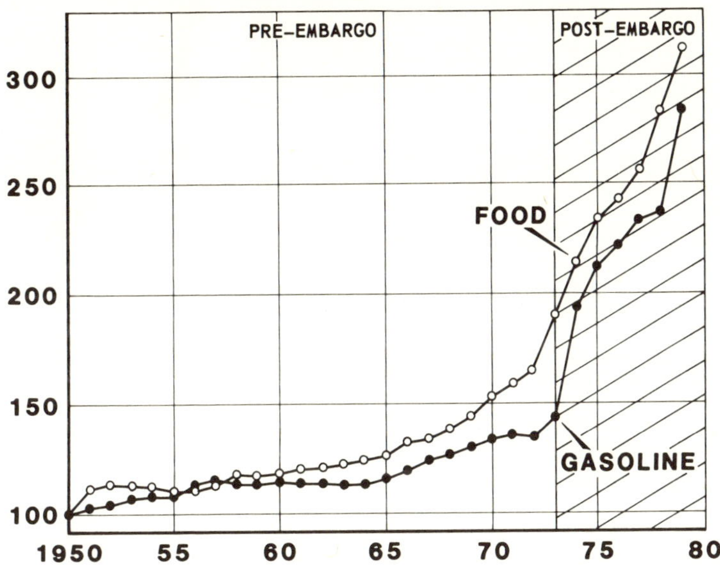

Figure 4-13

In comparing the price pattern of renewable resources with non-renewable ones, one would expect that developing scarcities of the non-renewable resource would drive up its prices well beyond those of the renewable resource. This was not the case of crude-oil-derived gasoline in the United States, where food prices rose much more rapidly than gasoline prices. For example, using 1950 as a base year for purposes of price comparisons, food prices had nearly doubled by 1973 (index 189.8), while the price of gasoline had risen by 44.8%, exactly half as much as food, After the embargo, gasoline prices did, of course, rise rapidly, but so did food prices; so much so, in fact, that by April of 1979 the non-renewable resource was still a bargain in the United States compared with food. This is not an accomplishment. It is an indictment of price controls, since the unnaturally low price of gasoline, due to price controls on crude oil, prevents the development of new energy resources and therefore keeps the United States dependent on imports.

Source: Gasoline Prices, See Fig. 5-12.
Food Prices, Calculated from *Handbook of Labor Statistics*, U.S. Dept. of Labor.

CHAPTER 5

MISCONCEPTIONS

Crude oil and natural gas come from beneath the ground. As to that, there seems no obvious disagreement among the energy experts who have multiplied in the halls of Congress and in the warrens of the bureaucracy. Yet that is where agreement ends — and the political myths begin. If reserves were as plentiful as misconceptions about the oil and gas industry, the energy crisis would be on its way to solution.

Some of the larger misconceptions should be addressed.

1. That higher prices do not translate into a higher drilling rate.

The assumption has been a popular one, as it affords excuses for keeping a federal ceiling on the price of crude oil and natural gas. If higher prices will not produce more oil and gas, why line the producers' pockets with unneeded revenues?

Not only common sense but experience stands the assumption on its head. As Adam Smith, in *The Wealth of Nations*, so acutely observes: "It is only for the sake of profit that any man employs a capital in the support of industry; and he will always, therefore, endeavour to employ it in the support of that industry of which the produce is likely to be of the greatest value, or to exchange for the greatest quantity either of money or of other goods." A higher price calls forth all the exertions of an entrepreneur. That is the marketplace. That is human nature.

So is it direct, observable experience. Consider natural gas. In the period 1968 to 1974, total gas completions in Sutton and Edwards counties, Texas, rose from ten to 240. There was no mystery about it. In that time span, the price paid producers rose from 14 cents per mcf to $1.40. Nor was the trend in any sense confined to Texas. In Eastern Ohio, total gas completions were below 200 in 1968; given price rises from 24.9 cents per mcf to $1.05, the completion rate by 1974 had increased seven-fold.

Of course not even high prices make drilling anything less than a risky business — one in which dry holes easily outnumber the successes. All the more reason, even so, to employ the incentive of higher prices. The more that is to be gained from a drilling project, the likelier it is to be hazarded. If the rewards of the search are comparatively small, only a few will choose to search; the higher reward brings into the game so many searchers that the odds of discovery are greatly multiplied.

2. That, however tempting the price, however hungry the contractor to drill, there is too little drilling equipment to go around.

Here again the law of supply and demand takes charge of the situation. The rise of the real price of oil and gas during the early '70s was accompanied by a doubling of drilling expenditures between 1971 and 1975. Between 1971 and early 1977, the number of rigs working in the Continental United States almost doubled — from 975 to 1,875. In a still shorter time span, 1973 to 1977, employment in the petroleum equipment supply industry soared from 25,000 to 62,500. So fast had the industry snapped back from the doldrums of the '60s that it was able to produce 679 drilling rigs a year — without an extra dollar's worth of capital spending.

3. That oil companies have been hoarding natural gas.

If the companies have been hoarding gas, they have hoarded it well: the minutest, most conscientious investigation has not turned up the missing gas. By way of example, several accusations of withholding were examined during the 94th Congress by the House Commerce Committee's Oversight and Investigations Subcommittee. The facts brought out refused stubbornly to square with the rhetoric expended.

In one case, Cities Service Co. was accused by the subcommittee staff of shutting down for pipe repairs its off-shore South Brazos Tract, A-76, not just to make repairs but also to increase the pressure for higher prices. In fact, gas from the tract was already under a 20-year contract at 23-29¢ per mcf at the time of the shut-down. Nor was there any more meat in the contention that Cities Service had dragged its feet in making the repairs. The workover rig which the staff said was readily available to Cities Service was, as testimony showed, too large to fit the drilling platform.

In another instance, Mobil Oil Corporation was faulted for intentionally failing to apply quickly enough for and accept a Federal Power Commission certificate allowing the sale of its gas from the Grand Isle 95 field. As the subcommittee staff saw it, Mobil had been in no hurry for certification. Another clearcut case of withholding!

Yet the facts point in quite a different direction. Not only did Mobil apply for an FPC certificate on March 10, 1975, six days after contracting to sell its gas — it urged prompt action by the commission so that a pipeline could be built right away and the gas delivered during the coming winter. Upon receipt of Mobil's urgent request, the FPC took no action at all for four months — and then merely asked for more information, a type of information it had never previously sought on any other application for FPC certification. Mobil supplied the data within a week, reemphasizing the need for haste. But the company's entreaties failed to end the wrangling. Earnestly as Mobil might plead for the granting of a certificate before September 5 — the last day that pipeline construction could begin — the FPC waited until after the cutoff date to respond. In responding, it turned down a contract provision fixing a 10-year term for the sales. Two months later, the FPC reversed itself; Mobil could have the 10-year term after all. Of course, by then it was far too late to start the gas flowing in time for winter. It was winter already.

So it went with the like accusations of Interior Secretary Cecil Andrus, who, upon taking office in 1977, ordered a sweeping inquiry into reports of withholding in the Gulf of Mexico fields. The *Oil & Gas Journal*, a publication based in Tulsa, Okla., found the operators of the five leases in question had never been asked by government investigators why they were not producing to the fullest. All told the *Journal* that, where economically feasible, they meant to produce as much as they could. One, Ocean Drilling & Exploration Co., said its reserves were too small and widespread — even if prices had been right, which they weren't — to warrant the installation of a pipeline. "For years," said a company spokesman, "we tried to get a pipeline to come in and take the gas."

It would be quite dangerous, of course, to assert that never has gas been withheld from production for reasons of self-interest. Yet two observations are pertinent. The first is that nothing has come of the entire withholding brouhaha. If withholding for the purpose of driving up gas prices had proved genuinely widespread, lawsuits and acts of Congress would have put a stop to it. In fact the purveyors of the accusations ended up retreating from them in some embarrassment. The accusations as a body were not made to stick. In them was the minimum of substance, the maximum of inflammatory political rhetoric.

The second observation is that expressions of economic self-interest are splendid guides to the realities of the marketplace. As we have seen, the traders and merchants of the Emperor Diocletian's day made known, by withholding their goods from market, that the emperor's price controllers were denying them a fair return on their investment. There was no reason whatever that they should sell their goods at prices too low to afford a profit. Had they done so, they would soon have gone out of business. And how would this have strengthened the imperial economy or aided the consumer? By withholding the fruits of their labor on grounds of self-interest, they signaled the emperor that his economic program was a failure. As, of course, it was. Witness simply the resounding crash with which it finally fell into ruins. Withholding is not merely an act of common sense; it is a kind of public service.

To the extent that deliberate "withholding" has existed in the natural gas industry, it has been withholding of a different character than that alleged by the federal investigators. It was not that producers as a class cheated on their commitments to the interstate market; it was that, insofar as they could, they avoided the interstate market entirely, selling instead to the intrastate market where they could receive a fair price for their product. We have seen how the intrastate market attracted larger and larger shares of the new gas being developed in the 1970s. But of course! Is it expected that men, whether they sell cigarettes, undershirts, or gas, will settle for a lower price if a higher one is available? For the plight of the interstate natural gas market, producers were not responsible; the federal government was.

4. That there is precious little gas yet to be found.

No energy crisis myth has less substance. To be sure, no one, whether gas producer or Department of Energy official, can count reserves as yet untouched by the drill. And yet the knowledgeable consensus is that plenty of gas remains unprospected — more gas indeed, than has been found up to the present day.

The Pitts Energy Group of Dallas calculates that 98 per cent of the nation's prospective gas-bearing sediments remain undrilled: three million square miles, to be precise.

Some two dozen reports by government or private groups have since 1963 established the abundance of undiscovered gas. A report, for example, by 150 geologists and professional engineers making up the Potential Gas Committee (under auspices of the Colorado School of Mines) projects the undiscovered potential gas supply at 923 to 973 trillion cubic feet. It is four times the nation's present proved reserves, twice the amount of gas produced since the 19th century.

Another report achieved a curious kind of celebrity in the pages of *The Wall Street Journal.* A task force of the Energy Research and Development Administration went to work in 1977 on a projection of gas supplies. Their report, known as the Market Oriented Program Planning Study (MOPPS), estimated, as *The Journal* summed up, "that at an uncontrolled price of $2.25 (per mcf) the nation would be awash with natural gas. It would bring in 230 trillion cubic feet of what USGS (the United States Geological Survey) calls 'inferred reserves,' make economic the 285 trillion cubic feet of Devonian shale in Appalachia, the 600 trillion cubic feet of Western 'tight sands' and between 200 trillion and 300 trillion cubic feet of coal-seam methane." Not counting the 20,000 to 50,000 trillion cubic feet that could be procured from geopressured methane, onshore and offshore, in the Gulf region, were the price to reach $2.50 to $3.00 per mcf.

Alas, so affirmative a prediction, right or wrong, dismayed ERDA. It cancelled the MOPPS study and reassigned those who had participated in it. The Carter administration was able once more to wallow in the unmitigated gloom that prompted the President to demand the retention of gas price controls, even if at a higher level than before. *The Journal* noted archly how the energy crisis was turning into a vested interest, the President and the energy bureaucracy, and even some of the regulated energy companies, all striving for their own reasons to keep it alive.

5. That higher wellhead prices for natural gas "rip off" the consumer.

Hardly. If only because, in common sense terms, it is better to have gas than not to have it. But there is even more to the matter than this. As Prof. (now U.S. Rep.) W. Phillip Gramm told the Senate Energy Committee in 1977, "Wellhead prices of natural gas represent only a fractional part of the price of natural gas paid at the point of consumption. This is especially significant in the Northeast, where 81 per cent of the residential gas bill is for pipeline and distribution costs. To the consumer of natural gas in the Northeast, a doubling of wellhead prices causes only a 20 per cent increase in point-of-use cost."

And what if, because of inadequate wellhead prices, there is too little gas to fill the pipelines? For the consumer there is misery in such a prospect. The supply gap must be closed in some manner. A 1977 report by economic consultants Zinder and Associates, in projecting the consequences of a 52-cent price ceiling for gas, foresaw in 1980 a 45 per cent shortage for the interstate gas pipelines. The shortage, the report said, would have to be made up by electricity, fuel oil, synthetic gas and imported liquefied gas at $4 per mcf.

The consumer pays yet another penalty when too little gas is available to fill the pipelines. The pipeline company is guaranteed a fixed rate of return on invested capital. Full, half-full, or empty, its facilities must be amortized. For smaller amounts of gas handled, the pipelines must accordingly charge consumers higher prices to make up the lost revenue. When in 1975 the pipelines serving the Northeast fell 15 per cent short of capacity — on account of the shrunken gas supply caused by controls — consumers paid an extra $1 billion. But what of those unable to buy gas at any price? The substitute fuels they were obliged to purchase, by way of making up for the 15 per cent shortage, cost as much as did the 85 per cent supply actually delivered!

CHAPTER 6

WINDFALL PROFITS

According to William Safire, it was President Franklin Roosevelt who incorporated the word "windfall," meaning an unexpected gift, into the language of politics.[1] Until then, the word had connoted luck or good fortune. In Roosevelt's mouth it became denunciatory; nor have matters changed much in the succeeding 40 years. What could be unluckier today than to reap a "windfall profit" before the eyes of a certain vocal species of politician?

Windfall profits — and how to relieve the oil industry of them — have become one of the burning issues of the energy crisis. More burning indeed than the more fundamental issue of how to overcome a scarcity of oil.

As Safire notes, the more formal title for "windfall profits tax" is "excess profits tax" — a phrase with origins dating back before the First World War. The current windfall profits debate is as old as public recognition of the energy shortage — which is to say that it dates back to late 1973.

To a Senate bill of that year conferring on the President special powers to reduce energy consumption, the House added a windfall profits tax. It was the contention of Rep. Jerome R. Waldie that the oil industry should make the same sacrifices as its customers and not persist in "its selfish, greedy and unpatriotic pursuit of the almighty dollar at the expense of the people, their health, their comfort, and their jobs." Senate conferees were able ultimately to get the provision dropped, but it was replaced with instructions that the President roll back the price of domestic oil exempted from price controls. President Nixon's own proposal for a windfall profits tax (10 per cent to 85 per cent on price increases of more than 50 cents) had fallen on deaf ears. For this and other compelling reasons he vetoed the emergency energy bill on March 6, 1974. The Senate upheld the veto.

[1]Safire's Political Dictionary, 1978 Edition (Random House-New York), p. 794.

Thereafter, the roar of windfall tax proposals dropped to a whisper. The nation having assimilated the quadrupling of oil prices, the issue of oil company profits had ceased to enrage, or even titillate. It was Jimmy Carter who revived the debate with his pledge, in April, 1979, to tax the oil companies' "huge and undeserved windfall profits" — this, by way of minimizing the presumptive harm that his plan to decontrol crude oil prices would wreak upon consumers. The President's rhetoric was strident and angry. He hurled scorn at the oil companies for daring to suppose they might collect revenues they had not "earned." "I will demand," said Carter — demand! — "that they use their new income to develop energy for America, and not to buy department stores and hotels, as some have done in the past." He warned that, "as surely as the sun will rise, the oil companies can be expected to keep the profits which they have not earned."

In political terms, if no other, it was an opportune time for attacking the oil companies. The price of gasoline was on the rise, owing to the temporary stoppage of Iranian oil and to a large price increase voted by OPEC in March, 1979 (Another, sharper increase was to follow on June 28, lifting oil prices more than 50 per cent higher than they had been in January.)

The windfall profits bill quickly passed the House, which voted to confiscate, through the late 1980s, $277 billion in oil company revenues. The Senate approved a lower tax — $178 billion. Whereupon a conference committee representing both houses settled on a compromise figure of $227 billion.

As finally passed, the windfall profits tax stood as high as 70% on certain types of crude oil. At those rates, given royalties and additional taxes, a one-dollar increase in domestic crude-oil prices generates a paltry 11 cents in after-tax profits to the oil companies. This is less than the amount accruing either to royalty owners or to the tax authorities. (See Table 1)

Table 1
EFFECT ON NET INCOME OF WINDFALL PROFITS TAX FOR A ONE-DOLLAR INCREASE IN THE PRICE OF CRUDE OIL

	No Windfall Tax	With Windfall Tax
Price Increase	$1.00	$1.00
Less Royalty	.14	.14
	.86	.86
70% Windfall Tax	0	.60
	.86	.26
Less Severance Tax	.05	.05
	.81	.21
Less 4% State Tax	.03	.01
	.78	.20
Less 45% Corp. Inc. Tax	.35	.09
Net Profit to Producers	.43	.11

Source: Research Reports, American Institute for Economic Research, Vol. XLVII, No. 3, January 21, 1980. Slight variation by authors to reflect 70% windfall profits tax.

Often overlooked is that the crude-oil tax scheme is an administrative abomination. The tax bill creates at least 20 different kinds of crude oil, each with its own price regulations. There is lower-tier oil, base-production-control-level upper-tier oil, upper-tier oil produced by major oil companies, upper-tier oil produced by independents; there is stripper oil for majors and independents, marginal oil, heavy oil, newly discovered oil, state royalty oil, Indian royalty oil, charitable hospital and school royalty oil, federal royalty oil, to name a few. If that sounds confusing, it is simplicity personified, compared to the 30 or so (no one knows for sure, how many) categories of natural gas, as defined by the Natural Gas Policy Act of 1978.

Not only this, but the windfall profits tax also takes into consideration the historical prices of the bulk of the oil under production, with the result that the per-barrel taxes vary widely from property to property, for given oil classifications. Thus, there will be literally tens of thousands of de-facto regulations covering this nation's crude-oil production, with individual base prices changing at monthly intervals.

A point worth noting early on is that the term "windfall profits" is the grossest misnomer, given the kind of tax actually levied. A true windfall profits tax would be self-liquidating; which is to say that it would expire eventually, independent of any action by Congress, once windfall profits themselves expired. Expiration could come about in two ways. First, through the depletion of "lower-tier" oil supplies, those whose prices are due to rise to world levels between now and 1985. Lower-tier oil is, in truth, disappearing already. In 1975, it averaged 62 per cent of domestic production; by April, 1979, it had declined to 34 per cent. With half the tax base forever gone, half the tax revenue is lost already, and the remainder continues to decline. The second reason that windfall profits are self-liquidating is that these profits are the primary source of capital needed for the development of alternative — and more expensive — energy sources. Among these sources are alcohol, shale oil and synthetic crude from coal, selling for an estimated $25 to $30 per equivalent barrel of crude oil. Of course the capital costs of developing such sources are immense. The Oil Shale Corporation (Tosco), a private consortium, has estimated the cost of a 50,000-barrel-a-day shale oil plant at $1 billion in 1977. For the financing of such ventures, "windfall profits" are indispensable. "Old oil," decontrolled and allowed to reach $26 a barrel, would yield a windfall profit of about $20 — enough to capitalize the development of substantial supplies of alternate fuel. By the time oil supplies were substantially depleted, and the U.S. economy switched over to alternative forms of energy, no windfall profits would remain. They would have been liquidated.

Not so with "windfall profits" as conceived by the Congress of the United States. It is not profits that Congress means to tax; it is price, which makes the proposed levy an excise tax like those applied to tobacco products, alcoholic beverages, gasoline, cosmetics, and so forth. An excise tax does not vanish when profits disappear; it endures until specifically repealed. Were the nation to convert to shale oil or synthetic fuel at a taxfree price of, say, $30 a barrel, a true windfall profits scheme would yield a true price of $30. Yet an excise tax of

$7 barrel would raise that price to $37.

The case against "windfall profits" taxes, in their spurious sense, is even more damaging than this. The case for such taxes fails on grounds of elementary fairness. In 1976, i.e., long before the enactment of any "windfall profits" tax, the top 10 oil companies were already paying an inordinate amount of taxes, as can be seen in Table 2.

Table 2
GOVERNMENT'S AND STOCKHOLDERS' INCOMES:
TOP TEN U.S. OIL COMPANIES
(1976)

Company	Dividends Paid	Corporate Taxes
EXXON	$1,220 million	$10,513 million
Texaco	543	1,430
Mobil	364	5,617
Standard of California	365	1,479
Gulf	336	1,823
Standard of Indiana	338	1,713
Shell (U.S.A.)	150	781
Atlantic Richfield	174	684
Continental Oil	120	964
Phillips Petroleum	134	713
	Total $3,744	$25,717

Inflammatory rhetoric to the contrary not withstanding, dividends paid out and taxes paid out *are* a proper basis of comparison, and on that basis the shareholder lost in 1976 with a score of 4 to 26. Retained earnings are, of course, overwhelmingly reinvested, at a risk, for a repetition of the losing game. Which makes one wonder why all the oil companies are not using all of their profits to get totally out of the game by investing in other industries.

What is a windfall, after all? A piece of unexpected good fortune, as we have seen. Now there are all sorts of windfalls, as in the pre-1980 real estate market, where inflation had constricted the housing market, just when products of the postwar baby boom were moving from beneath their families' roofs, seeking shelter of their own. The shortage of desirable real estate drove up prices. It was not uncommon for a home to have doubled in value in just three or four years. When such a home was sold, its owner quite clearly reaped a windfall profit. But the government did not propose taxing away that profit for social reasons, distributing the proceeds to apartment dwellers or young homebuyers cramped by stratospheric prices and interest rates.

Other such examples can be cited in the tumultuous economy of the '70s and '80s. Not the least of those examples is that of the federal government itself!

Consider the income tax. The existing rates were framed at a time of lower inflation — when, say, $20,000 a year was a generous income. But as Henny Youngman, the comedian, has so aptly expressed it, "I can't afford to live anymore on the income I used to dream about." Neither can the taxpayer afford the increasingly higher rates foisted on him by the inflexibility of federal tax law. The tax laws presuppose the affluence of $20,000-a-year wage earners. But $20,000 a year has become little more than an average income.

Moreover, inflation routinely shoves Americans into higher and higher tax brackets. The raises they receive to offset inflation not only fail generally to achieve their purpose — they confer on the government windfall tax revenues. No tax increase has been legislated; none has had to be. Inflation has done the job — by forcing up wages, it progressively forces up taxes. Prof. Milton Friedman's estimate is that in the single year, 1973, the government's windfall from inflation exceeded $25 billion. Why then does not the government levy a kind of "windfall profits tax" on *itself* — by legislating a tax cut?

There are, in fine, all manner of windfalls, the oil companies' being just one. As former Treasury Secretary William E. Simon has asked, "If windfall profits exist throughout the economy either in undiluted form or . . . as an element in almost all incomes, what possible justification is there for singling out a particular class of taxpayer and exacting a levy from them alone? This is discriminatory and unjust . . ."

The irony of the federal government's indignation at "windfall profits" grows all the more pointed when the oil tax itself is considered. Windfall revenues from inflation are nothing compared to those the government would reap under the proposed tax plans. To a government unable even with inflation revenues to balance its budget, the idea of tapping the oil companies' till comes as a godsend. What splendid projects these extra billions could finance.

Addressing the National Association of Counties on July 16, 1979, President Carter proposed "an investment of $140 billion for American energy security . . ." Where would the money come from? "All of this investment of federal funds must come from the windfall profits tax on the oil industry." A few minutes later, the President proposed $16.5 billion for mass transit and a tripling of federal funds to help the poor pay their energy bills. Again, windfall profits taxes would provide the money.

How spendidly convenient. The federal government's cash needs are staggering; the oil companies' supply of cash, thanks to price decontrol, happens to be large. What the government needs, it can pluck from the companies.

Nor is the government's tax war restricted to raids on corporate treasure chests. The imposition of a $4.62-a-barrel import fee on oil, collected in trust by the oil companies, is a frontal attack on the taxpaying public. A complicated and administratively heavy-handed entitlements program — the second entitlements program on the U.S. energy policy scene — assures (or attempts to assure) that this import fee is passed on exclusively to the gasoline-consuming public at the rate of 10 cents per gallon. That is over and above the windfall profits tax, which is more evenly applied to *all* crude-oil products — not just gasoline.

Issued through Presidential proclamation and made retroactive by some 3 weeks to March 15, 1980, the import fee will net the federal government some $14 billion per year if imports do not rise. A very big, very heroic and somewhat unlikely "if." Add to the import fee the $23 billion annual price tag of the windfall profits tax, and the government's total take comes out at $37 billion.

All discussions of "windfall profits" taxes and import fees occasion much high-mindness on the taxers' part. The money thus confiscated would benefit "the people". Again and again the refrain is heard. The notion, however, is a dubious one.

To begin with, taxes are not borne by companies but by consumers. Taxes paid are a cost of doing business; the higher the tax, the lower the profitability of the business in question. Of course a company's ability to pass on its costs through higher prices is limited by competition — and, in the oil companies' case, by federal controls. Yet the fact remains: low taxes strengthen a company; afford it capital it would otherwise have to hand over to the government; keep profits healthy and prices competitive; attract new investors; permit the purchase of new equipment, the building of new facilities, the hiring of new workers. Low taxes, in short, encourage production, which means greater availability of products, which in turn means low — or at the very least, stable — prices. Has the public no interest in stable prices? You might suppose not, to hear the windfall taxers go on.

What is the meaning of these observations in the context of oil? The meaning is simply that by vastly increasing the oil industry's tax bill, we impair the industry's ability to expand production of domestic oil and thus to blunt OPEC's "oil weapon."

William E. Simon writes: "The faster profits rise, the faster new oil will come into market and the faster OPEC's stranglehold will be removed. A tax will hinder this beneficial process, either slowing down energy independence or making the consumer pay more for it. In other words, it is Big Government which is ripping us off, not Big Oil."

The common sense of Simon's words can easily be seen. Drilling costs money; the more money the oil companies and the independents have, the more wells they can drill; the fewer wells drilled, the farther off the day that the United States achieves so much as a measure of energy independence.

The problem is compounded by the abolition of the oil depletion allowance for major producers. The allowance — a tax deduction conferred by Congress in 1926 to compensate oilmen for the non-renewable nature of their product — is enjoyed by all the other major mineral industries. However, such was the political outcry during the 1960s and 1970s against the oil industry that the allowance was repealed, except for smaller producers. To the oil industry, depletion represented a ready source of capital for reinvestment; its passing left profits as the only internal source of financing. Windfall profits tax or no, oil profits are severely strained by the intense need for more funds.

The oil companies' critics respond that none of this matters. Why, the companies are as rich as Croesus; one has only to look at their profits reports, reflecting in 1979 gains of 50, 75, 100 per cent over 1978. The figures, which result from the major price increases implemented by the Organization of Petroleum Exporting Countries, are deceptive in the extreme. The large percentage increases that make such lively reading on the evening news are mostly generated from foreign operations unaffected by American price controls.

In any case, the companies' capital spending needs are immense. For the top 27 major oil companies, capital expenditures have more than doubled between 1973 and 1978, from almost $15 billion to nearly $30 billion.[1] From each dollar earned in 1978, 77.9 cents went to operating costs, which dwarfed the next-largest source of spending, taxes, at 11.5 cents; profits were a less-than-colossal 4.3 cents; 2.4 cents of these profits were reinvested, leaving 1.9 cents for dividends, payable to the real owners of the oil companies — their millions of shareholders, far from all of them affluent, many of them elderly, retired, or both.

Overall, the major oil companies' rate of return on investment seems pleasing enough — 13 to 15 per cent, compared with 9 to 13 per cent in the pre-embargo years. But consider the rates of other manufacturing industries prior to the 1979-80 credit crunch: housing construction and coal

[1]These companies are: Amerada Hess Corp.; Ashland Oil, Inc.; Atlantic Richfield; The British Petroleum Co. Ltd.; Champlin Petroleum; Cities Service; Clark Oil & Refining Corp.; Compagnie Francaise des Petroles; Continental Oil; Exxon; Getty Oil; Gulf Oil; The Louisiana Land and Exploration Co.; Marathon Oil; Mobil Oil; Murphy Oil Corp.; Petrofina Societe Anonyme; Phillips Petroleum; Royal Dutch/Shell; Standard Oil of California; Standard Oil (Indiana); The Standard Oil Co. (Ohio); Sun Company; The Superior Oil Co.; Texaco; Tosco Corp.; and Union Oil of California.

For a more detailed discussion of the financial performance of these companies, see Appendix to this chapter.

mining, over 30 per cent; more than 20 per cent for the manufacturers of semi-conductors, household furniture, nuts and bolts, radio and TV sets, and motorcycles and bicycle parts. It may be that the oil industry is large and wealthy. Yet it is not nearly so wealthy as its critics in Congress and the media enjoy pretending. It will be still less wealthy now that the first bill for windfall profits taxes has been presented.

The tax is clearly discriminatory, as clearly it was meant to be. Yet the message it conveys is not just for the oil industry but for American business as a whole. The message is that profit is all right — but not in disagreeably large doses. Further, that private ownership of natural resources is acceptable — except when it conflicts with public goals arbitrarily established by Congress and the White House. The message is not an encouraging one for those millions who risk their capital in the expectation of a better life for themselves and their families.

It has been argued by the windfall taxers that there is no choice. Let the companies keep the money and they would use it to drill for more oil instead of developing solar energy and synthetic fuels. The argument assumes some kind of superior foresight and economic understanding on the part of government. Likewise it fails to account for the fact that the oil companies already are engaged in the search for alternative energy sources. As soon as it became plain that an energy crises was in the offing, the companies, which did not grow big through stupidity, began buying coal supplies. At the right price, coal can profitably be turned into gas. The companies also stepped up efforts to extract oil from shale — a costly undertaking, again justifiable only at the right price. The workings of the free marketplace, not the deliberations of a presidential speech-writing team, is what causes such purely economic decisions to be made.

Of course it only makes sense, given America's current dependence on oil, that the companies should try to find more of it — if only to smooth the country's path to the development farther down the road of more exotic energy sources. The White House is not unwilling that the companies should drill *some* wells. Indeed, it asserts that, even with the stiff windfall profits tax, they will have enough money to do so. A bold assertion, surely. How much is enough? The White House cannot know, nor can the Congress. "Enough" is whatever amount is indicated by the cost of drilling the most attractive prospects. At all events, the "windfall profits" tax was not framed with oil production in mind. The administration did not ask how much money the companies needed. It asked how much the government needed to expand mass transit, develop synthetic fuels, and so on. Whatever amount was left over was what the White House was willing for the companies to keep. It was not even contemplated that the industry could exempt itself from confiscatory taxation by plowing its new revenues back into the search for oil. The White House wanted money. It saw the oil companies' coffers brimming over. It ordered the companies to stand and deliver. The matter is as simple — and as tragic — as that.

APPENDIX TO CHAPTER 6 WINDFALL PROFITS ON GRAPHS

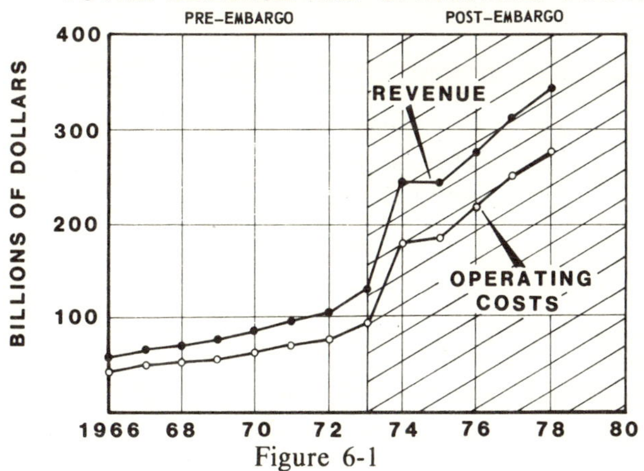

Figure 6-1

Measured in terms of total revenue, the oil embargo practically doubled the size of the oil industry overnight. This was due almost exclusively to the increase in the price of imported crude oil. During the pre-embargo period (1966 to 1972), the group's total revenue rose from \$59.6 billion to \$106.3 billion an average of 10% per year. By 1978, total revenues had risen to \$346.9 billion. This corresponds to a post-embargo growth rate of nearly 22% per year. People often have the impression that oil revenues rose while costs remained the same. That is a mistaken view. During the pre-embargo period (1966 to 1972), direct operating costs grew at essentially identical rates as revenues. After the embargo, costs rose *faster* than revenues. Running between 72 and 73 percent of total revenues prior to the embargo, direct operating costs had risen to 80% of total revenues by 1978. The group's total operating costs that year amounted to \$277 billion.

Source: *Financial Analysis of a Group of Petroleum Companies*, 1978; Chase Manhattan Bank.

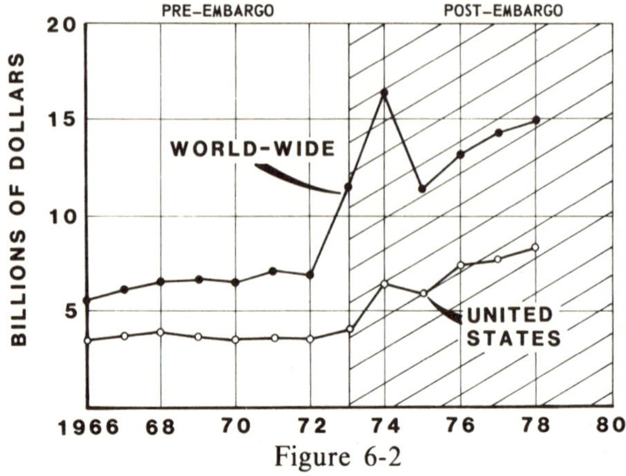

Figure 6-2

The oil industry is a multinational industry par excellence. The group of major oil companies here under consideration has extensive production, purchasing, refining and selling operations all over the globe. Accordingly, it has profits on both its domestic and foreign ventures. U.S. profits of the group remained essentially stable before the embargo at some 3.4 to 3.9 billion dollars. After inflation adjustment, U.S. profits have been declining over the period 1966 to 1972. During the post-embargo period (1973 to 1978), the majors' profits from U.S. operations rose, but not spectacularly. Rising from a 1973 level of $4.1 billion, the group's U.S. profits reached $8.3 billion in 1978. This corresponds to an annual growth rate in nominal profits of 15%. The restriction in U.S. profits was the result of price controls. Desirable as they might appear on the surface, these controls prevented the generation of capital that might have been used in developing alternative domestic energy resources. The so-called "obscene" post-embargo profits, especially those of 1974, came from the group's foreign operations. Overall world-wide profits that year were $16.4 billion, up from $6.9 billion two years earlier. By 1975, the transitional shock had passed and a smoother growth pattern emerged, leading to 1978 profits of $15.0 billion.

Source: *Financial Analysis of a Group of Petroleum Companies,* 1978; Chase Manhattan Bank.

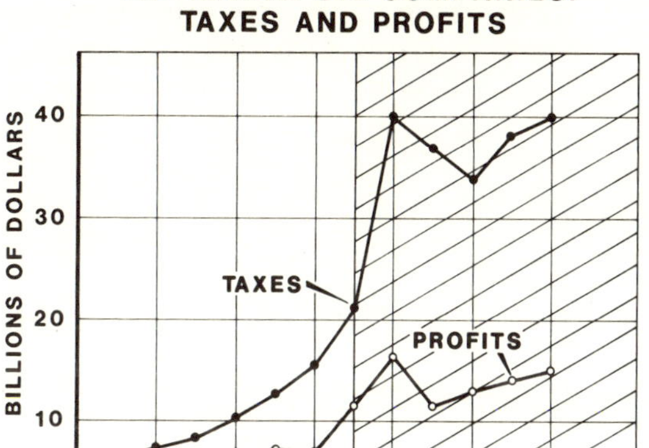

Figure 6-3:

Rightly or wrongly, some executives of the U.S. oil industry have developed guilt feelings about their companies' profits. The president of one of the group's companies has actually advocated price controls, with a plow-back provision for investments in energy exploration, because he feels that high profits in the oil industry established some sort of guilt by association with OPEC. Such a belief reveals a profound misunderstanding of the role of profits in a capitalistic society. Still, if such guilt feelings were appropriate, the government would have to feel approximately two-and-a-half times as guilty as our mistaken corporate president, since government taxes, roughly equal to profits in 1966 and 1967, rose from $5.5 billion in 1966 to nearly $40 billion in 1978. By contrast, the group's world-wide profits in 1978 were $15.0 billion. The taxes shown here do not even come close to representing the government's total tax bite on oil and natural gas operations. Not included are excise taxes on gasoline and other refined products that are levied directly on the consumer. In 1976, these amounted to $40.2 billion. And, of course, the profits are again taxed as personal income to the shareholders. In 1978, these taxes were roughly estimated at $6 billion, for an overall tax bite of $86 billion, compared to an after-personal-income-tax profit of $9 billion.

Source: *Financial Analysis of a Group of Petroleum Companies,* 1978; Chase Manhattan Bank.

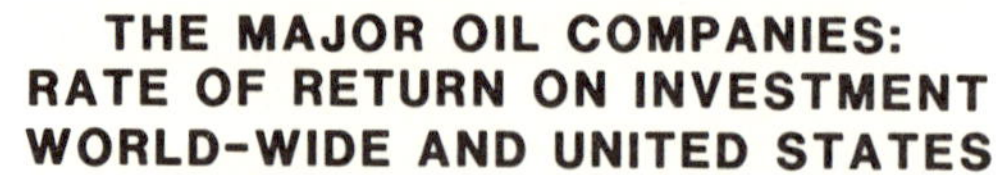

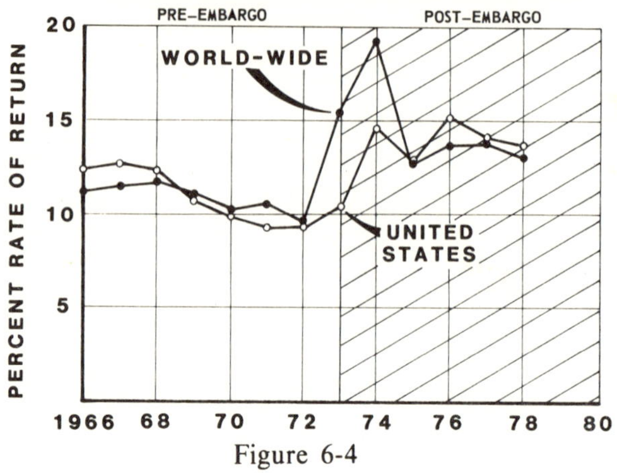

Figure 6-4

If they make any profits at all, large companies tend to make large profits, and small companies small profits — at least if measured in terms of total dollars. Major oil companies are large: they make large profits. The point here is that the dollar profits of oil companies, having received so much attention from all quarters, are meaningless and, worse, misleading. What matters is the *rate of return on investment,* generally expressed in percent. The group's pre-embargo rate of return on investment, both domestic and world-wide, ran between 9 and 13% — a somewhat mediocre performance compared with other industries. If anything, pre-embargo rates of return declined slightly over the period 1966 to 1972. The group's post-embargo rates of return are higher, but not spectacularly so. With the exception of the transitional year of 1974 (19% world-wide rate of return), the rates now fall into a range of 13 to 15%. For example, the domestic rate of return in 1978 was 13.8%, compared to more than 30% in bituminous coal and lignite mining or in housing construction, and to well over 20% in motorcycles and bicycles parts, semiconductors, household furniture, nuts and bolts, radio and TV sets, to name a few manufacturing industries.

Sources: *Financial Analysis of a Group of Petroleum Companies,* 1978, Chase Manhattan Bank.
Key Business Ratios, 1979, Dun & Bradstreet.

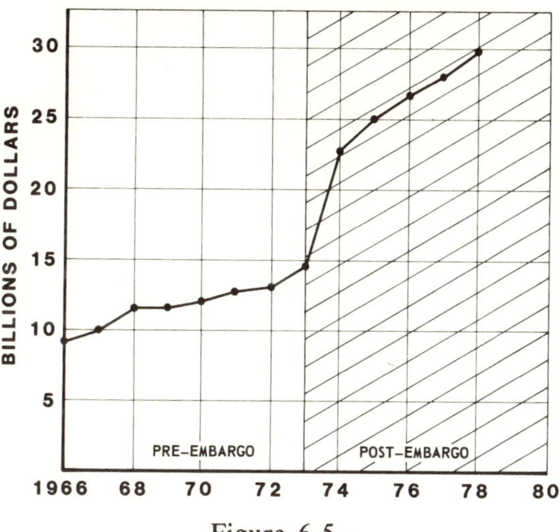

Figure 6-5

One of the least understood aspects of the energy dilemma is the enormous capital intensity of the oil industry as measured by the total capital investment per employee. Total capital investment is net worth plus long-term debt. By that measure, the majors have invested something like $120,000 per employee, 7½ times as much as the automotive industry's total capitalization of $16,000 per employee (1974-data). The group's annual capital expenditures rose steadily in the pre-embargo period, from $9.4 billion in 1966 to $14.6 billion in 1973. This corresponds to an annual growth rate of 6.5%. After the embargo, the group's capital expenditures rose rapidly, reaching $29.9 billion in 1978. The capital requirements for the development of conventional energy sources are staggering, but they are dwarfed by the capital that *will* be needed to develop non-conventional energy such as shale oil, tertiary oil, synthetic oil and gas from coal, and oil from tar sands. The need for substantial profits becomes inescapable in view of this nation's current and future energy related capital demands. If these profits are taxed away or eroded through price controls, the U.S. dependence on energy from foreign sources is bound to rise.

Source: *Financial Analysis of a Group of Petroleum Companies,* 1978, Chase Manhattan Bank.

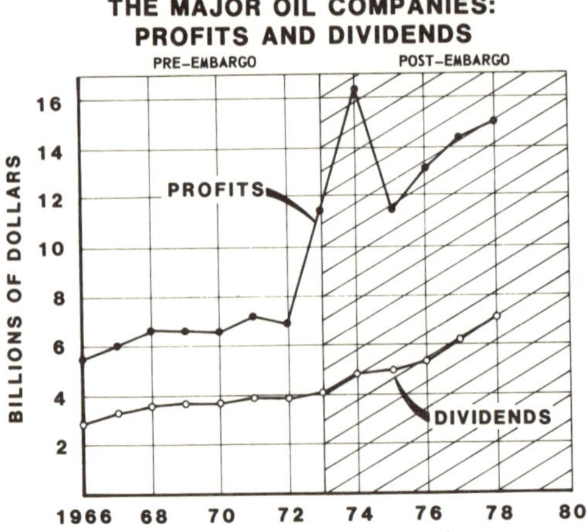

Figure 6-6

The primary internal source of capital, and the only significant one for the major oil companies, is profits. Write-offs such as depreciation, depletion, and amortization are a source of cash flows but not of capital; they are officially called "capital consumption allowances" in the U.S. national income accounts. By setting aside cash to replace worn-out capital equipment, these allowances merely perpetuate a corporation's capital stock, and less than perpetuate it if rampant inflation raises the replacement value of capital goods significantly above the original acquisition price. Thus, during inflationary periods, a corporation has to dip into profits just to perpetuate itself, since write-offs by themselves are insufficient in size to fulfill their intended function.

The group's profits in 1966 were $5.7 billion, of which $3.0 billion (or 53%) was paid out to the shareholders as dividends. The rest was re-invested. Rising at an annual rate roughly half the rate of inflation, the group's profits reached $6.9 billion in 1972. Dividends that year again were a little more than half the profits. All of the sudden profit increases of 1973 and 1974 were re-invested, with hardly any impact on dividend payments. Thus, nearly ¾ of the 1974 profits were re-invested. By 1978, profits were $15.0 billion, of which $7.0 billion were paid out as dividends, while $8.0 billion were re-invested.

Source: *Financial Analysis of a Group of Petroleum Companies,* 1978, Chase Manhattan Bank.

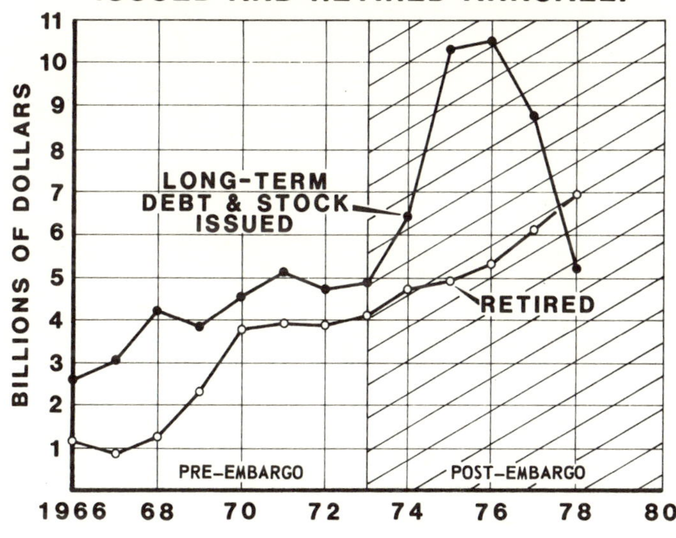

Figure 6-7

Outside sources of capital are long-term debt and new equity capital. These are acquired through the sale of bonds and stock, respectively. The major oil companies tapped such outside capital to meet their needs, but they dealt almost exclusively in bonds. Thus, more than 90% of the above graphs represent debt financing. During the pre-embargo period, the group issued more bonds than it retired, for a net capital infusion of around 1 to 3 billion dollars per year. The urgent need for alternative sources of crude oil (including Prudhoe Bay's reserves and the building of the Alyeska Pipeline) sent the group's capital needs soaring in 1974-1977. By 1978, with the Alyeska Pipeline Project completed and with a long-term debt structure beginning to be a real burden, new debt issues declined below the level of debt retirement. Of course, rapidly rising interest rates on corporate bonds compounded the group's debt-financing problems.

Source: *Financial Analysis of a Group of Petroleum Companies,* 1978, Chase Manhattan Bank.

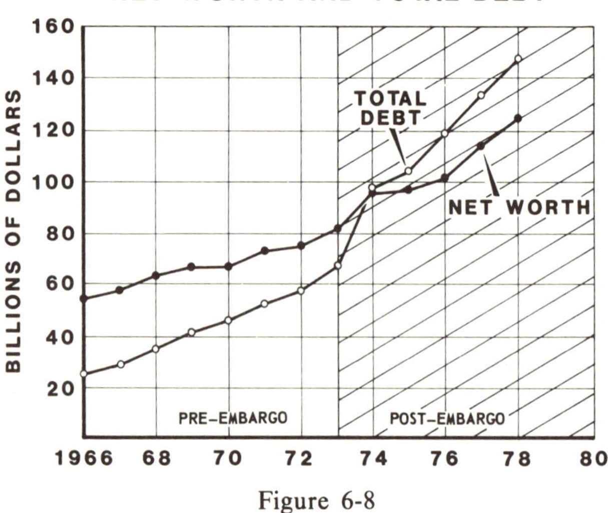

Figure 6-8

The total indebtedness of the major oil companies is a staggering $150 billion, and rising. That's roughly one-fifth of the U.S. national debt incurred by the federal government, whose spending activities involve 22¢ of every dollar spent in the United States. The group's total debt is all the more amazing if one considers that it was only $26.4 billion a mere 12 years earlier in 1966. The total indebtedness of the major oil companies has been rising so fast, both before and after the embargo, that it now exceeds the group's net worth. When that happens in a corporation, then (according to Dun & Bradstreet) "the equity of creditors in the assets of the business exceeds that of the owners", and the common joke about a bank's owning an entrepeneur's business takes on grim overtones. Be this as it may, over the period 1966 to 1978, when the group's indebtedness more than quintupled, its equity ownership only doubled — or increased by a factor of 2.3, to be precise. Starting in 1966 at $54.1 billion, the majors' net worth rose to $125.4 billion by 1978 — some $25 billion short of their total debt. Again, the only way to alleviate this problem is to permit profits to rise to their natural (uncontrolled) levels.

Source: *Financial Analysis of a Group of Petroleum Companies,* 1978, Chase Manhattan Bank.

THE MAJOR OIL COMPANIES: WHERE THE REVENUE DOLLAR WENT

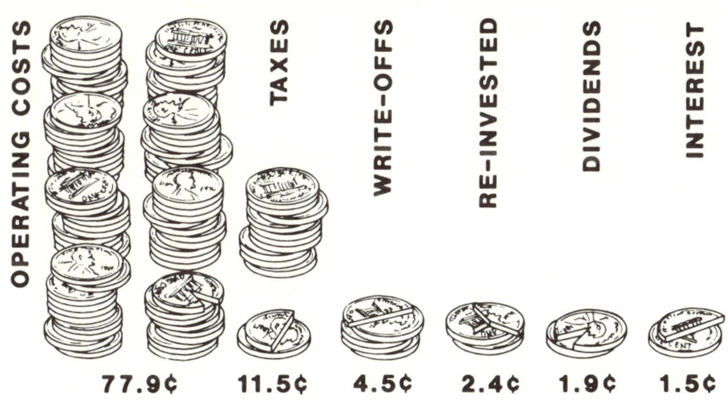

Figure 6-9

Soaring post-embargo revenues have been met, and more than met, by soaring costs. In 1978, fully 77.9 cents of every revenue dollar received by the major oil companies was spent on direct operating costs, such as acquisition of oil and natural gas, production, drilling and exploration, etc. Of the remaining 22.1 cents, more than half (11.5 cents, to be exact) was spent on taxes As mentioned earlier, this does not take into consideration excise taxes and individual income taxes on dividends. Write-offs represented 4.5 cents per revenue dollar. But again, due to inflation the write-offs are insufficient in size to fulfill their role as a capital replacement fund. The interest charge on the group's indebtedness took 1.5 cents out of every revenue dollar, leaving a grand total of 4.3 cents in net profits. Of this, 2.4 cents was retained by the group for purposes of re-investment, while 1.9 cents was paid out to shareholders in the form of dividends, subject to further taxation by various governmental authorities.

Source: *Financial Analysis of a Group of Petroleum Companies,* 1978, Chase Manhattan Bank.

CHAPTER 7

THOSE GASOLINE LINES

The autumn of 1973, the dark autumn of the energy crisis, brought a sight never before seen in America — automobiles stretched one after another across service station lots and down the street; waiting for, of all things, gasoline. The automobile was almost, next to the bald eagle, the American symbol — emblem of the nation's restless mobility, its eagerness to get on with the job at hand. Suddenly, for America's 130 million motor vehicles, there was too little gas to go around, and what gas was available sold for prices formerly unheard of.

The lines endured through the winter. Then they disappeared as OPEC oil began once more to flow to the refineries. In the spring of 1979, the lines suddenly sprang up again — longer, if anything, than before. American motorists played a game of hunt-the-station: find a gasoline retailer with gas to sell; never mind his prices, nor the duration of the wait, in lines that might stretch for blocks. America had wound up the '60s thrusting for the moon. It finished the '70s parked bumper-to-bumper.

For so startling a comedown there was, of course, one ready explanation: the nation was running out of oil; at any rate, oil that could be readily produced. But there was at least one subsidiary explanation that warranted a far closer look than was ordinarily accorded it. It was that the federal government, by its attempts to spread the shortage more evenly, had only made it worse.

A preliminary word should be said about the marketplace and its uncanny knack for allocating the right quantity of a given commodity to the right locations. *Price* is what guides the marketplace in its deliberations. A shortage of that universally popular commodity, widgets, arises, let us say, in New Mexico and Arizona. Now, because the two Southwestern states have too few widgets, their desire for them exceeds that of Pennsylvania and New York. The people of Arizona and New Mexico are willing to pay a higher price for whatever widget supplies can be procured. So widgets flow to them from the better-provisioned states, lured as by a magnet. If Arizona and New Mexico's need is vast, it may be that the supplies they pull away from the other states may cause shortages there. But here again, the price mechanism, like a superintending providence, acts to even things out. Those localities where the need is greatest bid for their own supplies of widgets. Again, prices rise most where the need is keenest.

In due course, the market provides a still greater service. The price incentives offered for widgets cause widget manufacturers to increase their output, or to bring in supplies from abroad, assuming full domestic capacity has temporarily been reached. Likewise, new manufacturers may enter the business and old ones expand their plants and hire new workers. The widget shortage becomes an *equilibrium*, if not a surplus. The sufficiency of supply, other things being equal, holds the prices level if it does not, in fact, force them down. So the marketplace works — when the marketplace is allowed to work. Which occasions grow fewer and fewer as the 20th century drags to its close.

In the midst of the energy crisis, the federal government took it upon itself — for reasons conceivably having as much to do with fair play as with raw politics — to determine where gasoline supplies should go. It was not necessarily a hard decision to understand, given the context. The nation was facing a drastic shortage of a basic component in its energy supply. Never in peacetime had there been so dramatic a shortage of so essential a commodity. Economic theory was all very well, but could it be trusted in an emergency? The right answer, of course, was *yes*; but it is possible to see how good men, in an hour of strain, could have said no.

Under authority of a law passed in 1973, a complex set of gasoline allocation formulas was administered by the Federal Energy Office. The FEO sought to provide each part of the country roughly the same quantity of gasoline, and to ensure that independent (that is, non-oil-company-affiliated) gasoline retailers remained in business. The allocations system was unfortunately — perhaps inevitably, given the lack of alternatives — based on historical consumption patterns. Where the cars had been in 1972, that was where the gasoline went in 1973, whether the cars were still there or not. And so it happened that urban gasoline stations ran quickly out of supplies, while stations in small towns and on the interstate highways were brimful of gasoline. In newly developed areas, where demand had been much smaller in 1972, gasoline lines were particularly long.

President Nixon's "energy czar" throughout the crisis was William E. Simon, a New York investment banker who would subsequently become secretary of the Treasury. In his piercing book on the slide toward collectivism, "A Time for Truth," Simon writes pungently of the days when all the good will — and power — in the world could not make gasoline go where it belonged:

> As for the centralized allocation process itself, the kindest thing I can say about it is that it was a disaster. Even with a stack of sensible-sounding plans for even-handed allocation all over the country, the system kept falling apart, and chunks of the populace suddenly found themselves without gas. There was no logic to the pattern of failures. In Palm Beach suddenly there was no gas, while 10 miles away gas was plentiful. Parts of New Jersey suddenly went dry, while other parts of New Jersey were well supplied. Every day, in different parts of the country, people waited in line for gasoline for two, three, and four hours. The normal market distribution system is so complex, yet so smooth that no government mechanism could simulate it. All we were actually doing with our so-called bureaucratic efficiency was damaging the existent distribution system. As the shortages grew more erratic and unpredictable, people began to 'top off' their tanks. Instead of waiting, as is customary, to refill the tank when it is about one-quarter full, all over the country people started buying 50 cents' worth of gas, a dollar's worth of gas, using every opportunity to keep their tanks full at all times.

And that fiercely compounded the shortages and expanded the queues. The psychology of hysteria took over.

Essentially the allocation plan had failed because there had been a ludicrous reliance on a little legion of government lawyers, who drafted their regulations in indecipherable language, and bureaucratic technocrats, who imagined that they could simulate the complex free-market processes by pushing computer buttons. In fact, they couldn't.

Individuals learn from previous mistakes and mend their ways. Governments, seemingly, never do. The fact that a given policy has worked poorly is, to those who administer it, no recommendation for killing it. Scapegoats can always be found, and in 1973-74, scapegoats were available in plenty — the "greedy" oil companies; the "avaricious" Arabs; the consumer himself, unquenchably thirsty for gasoline and therefore unwilling to sacrifice for the common good. That the culprit might be the makers and enforcers of federal policies was a proposition unlikely to occur to many of those same makers and enforcers.

And so when the traumas of the 1973-74 winter were behind, instead of scrapping controls on gasoline, the government retained them, as much for symbolic purposes as for any other reason. There — that would show the oil companies and the Arabs that a vigilant Congress and energy bureaucracy were on the job!

Yet a government agency's public face is not its only face. Behind the walls of the energy bureaucracy, invisible to the public gaze, there was arising a startling view of the energy crisis. It was that government controls were the proximate cause of the gasoline shortage. By 1979, the conviction had begun to harden. It was expressed in an internal memorandum by the Department of Energy, setting out the case against controls as lucidly and persuasively as any oil industry economist or oil-state senator has done.

The memorandum notes that gasoline price controls have "helped create a potential shortage of unleaded gasoline." The controls restrict selling price, including profit margin, to the price obtaining on May 15, 1973, plus cost increases. A refiner may pass on to consumer the higher cost of crude oil. But may he increase his rate of return on investment? Emphatically he may not.

"This inability to earn an adequate return on new capital investment, coupled with fixed profit margins," the memorandum argues, "acts as disincentive to needed capital investment. Although consumer demand for unleaded and high-octane gasoline has greatly increased in the past few years and continues to increase, that increased demand has not triggered normal price and investment changes. Refiners have not made investments in unleaded gasoline production facilities sufficient to meet the increased demand because the price controls prevent the profitability of that product from rising with the changing demand."

Besides this, the memorandum concluded that controls in general had "contributed to many inefficiencies in the market, inhibited experimentation with new pricing structures, and created serious distortions in the competitive relationship of firms."

For example, one major refiner staking out a new territory for its wholesale gasoline was caught with its prices down when controls were imposed. Accordingly, the refiner now services that region at artificially low prices. The Department of Energy memorandum also cites the case of a major marketer who continues to keep several previously unprofitable retail outlets open, despite long-range plans for their closure. Federal regulations enable him to do so, as they permit passing through to the consumer the cost of operating these inefficient stations.

The government's allocation controls have been no less unfortunate, the energy department memorandum makes clear. "In freezing supplier/purchaser relationships, the gasoline allocation controls have introduced a rigidity that is intolerable in the normally dynamic market process. In the six-year period since the allocation controls were imposed, gasoline marketing has experienced a number of changes. Self-serve islands, high-volume outlets, and convenience store/gas operations have been introduced, but development of these and other new marketing concepts has been impeded by the regulations."

The allocations program has "protected many inefficient firms that would not have survived in an uncontrolled market. The need, and even opportunity, to bargain for supply has been eliminated to a large extent. The allocation controls, together with price controls, have arbitrarily placed some firms in very favorable circumstances and others in untenable situations."

Well, cannot the allocations base period be moved up in time to reflect changed conditions? It can, of course. "A base period that is closer in time to the present will be more representative of current market conditions, but after six years of controls, those conditions will not approximate competitive market conditions and the beneficial results therein. A new base period just allows a different, but not necessarily better, set of arbitrary winners and losers."

In sum, "We are building distortions on top of distortions in petroleum industry markets that have been subject to 'temporary' controls longer than any counterpart in American economic history. Imposing regulations in the market introduces rigidities which cause innumerable things to go wrong with the process. We then try to correct these unanticipated defects by introducing new regulations, which cause more things to go wrong, so we introduce more regulations. But there is no hope of full correction because regulations try to freeze conditions that are inherently dynamic. Full correction of the defects can only be achieved by removing the regulations which have caused them."

For all the harm they inflict, gasoline allocation and price controls have not been the only forms of federal regulation under which refiners labor. The government's crude cost "entitlements" program was designed to equalize crude oil costs for all domestic refiners, whether they purchase high-or low-cost supplies. Such, at any rate, was its high-minded intention.

The entitlements program works thus: a refiner who processes a crude-oil unit priced higher than the national average is issued entitlements; refiners processing a lower-than-average crude-oil mix must buy these entitlements. The price, set by the Energy Department, reflects the price differential between the average cost of both imported and stripper oil and the average cost of old oil. Which differential presently is over $18 a barrel.

The cost to the purchasers of entitlements is considerable. The big loser under the program, Shell Oil Company, had paid $1.5 billion by August, 1979, followed by American Oil Company ($1.4 billion) and Gulf Oil Corp ($1.2 billion). The clear winner is Amerada Hess, which refines mostly imported oil; by the same date, it had received $1 billion in subsidies from its competitors. Next in line were Sohio and Ashland Oil, respectively.

The beneficiaries generally are refineries located in non-oil-producing areas and small refineries — those with a refinery capacity of less than 175,000 barrels a day. Naturally, the money handed over is money no longer available to the larger refiners for improvement or expansion of refinery facilities or for exploration and production of crude oil. That much worse, then, grows our crude oil and gasoline shortage.

All these programs of control and redistribution — of money or of supplies — were intended for the consumer's ultimate benefit. How well the consumer was served became apparent during the spring and summer of 1979, as gasoline lines once more appeared throughout the country — long, dark serpents roused from their hiding places, coiling anew around the nation's energy lifeline. The immediate cause of the lines was the total shutdown of Iranian oil exports amid the revolutionary turbulence accompanying the downfall and exile of Shah Mohammed Reza Pahlavi. By early summer, the lines had become critical. To find a station dispensing gasoline — far less to accomplish a purchase after waits of up to a couple of hours — became a major feat. The potentially productive time squandered while seeking a station or waiting in line has not been calculated, if indeed it could be. But it was immense. The governors of most states felt obliged to impose some kind of official rationing plan. In California, the plan affected the whole state; in Texas, merely the largest cities where supply problems were most acute. Slowly the lines disappeared but not the fundamental problem, which is that an excess of government control and regulation has fostered a shortage of the nation's economic lifeblood, energy.

If there had been doubt of this, yet another internal report by the Department of Energy went far toward clarifying the matter. DOE, whose perceptiveness regarding the energy crisis grows and grows, sought to explain the reemergence of the gasoline lines. It concluded — to the dismay of those congressmen and consumerists deeply committed to regulation — that the oil companies could not, by reasonable men, be held responsible. The blame had to be laid elsewhere.

First of all, said DOE, there occurred a large daily drop in domestic crude-oil production from 1978 to 1979. Second, oil stocks were drawn down from the exceedingly high levels to which they had been built up earlier in expectation of an OPEC price increase. Third, the United States received a smaller percentage of world oil than usual. Imports increased, but not enough.

Of old, said Doe, the nation could have offset an imports shortfall by opening wider its own wells. But that ability, of course, no longer existed. Indeed, "domestic crude oil available . . . averaged about 215,000 (barrels per day) less than in the same period in 1978. The data reflect the continuation of the basic trend of decline in crude-oil production in the lower 48 states." Accordingly, from February through May, U.S. oil supplies averaged 170,000 barrels a day below the previous year; the gap would have been larger but for a nominal increase in net daily imports which rose, overall, despite the Iranian shutdown.

The rumor mills that spring and summer were busy as always, grinding out far-fetched tales of gasoline tankers waiting offshore for prices to rise; of fully loaded gasoline trucks secretly spewing out their contents in the desert. All of which was so much eyewash. As the energy department report makes clear, the shortage was not caused by hoarding. The nation faced a real and severe scarcity of energy. In May, refiners drew down their crude oil stocks by 6 million barrels.

And what of controls and allocations? The Department of Energy report was commendably unsparing in its notation of these as contributing factors to the gasoline lines. With gasoline prices firmly under control, crude oil supplies that might have ended up as gasoline presumably went elsewhere in search of greater economic opportunity. This is what the authors of the report suggest. They inflict large bruises, too, on the allocations program, noting its inability to coordinate present supplies with past demand. "A historical demand base period inherently cannot reflect the most recent changes in demand for gasoline . . . Any allocation system is likely to be ineffective in responding quickly to continuing changes in demand."

Nor was this all the government accomplished through its tender solicitude for the consumer. By establishing a system of "priority" use for certain kinds of fuels, it encouraged a scramble for such supplies. This led in turn to serious abuses. The government also discouraged the purchase of oil on the spot market. Prices in this market were sharply higher, to be sure; but, when there are long gasoline lines in every neighborhood, fuel is fuel.

If only the report could have probed to the bottom of the matter, singling out control of crude oil prices as the fundamental difficulty. It refrained; but so broad is its indictment of government policy and meddling in the energy market that no one need complain. Except, naturally, fellow bureaucrats far more committed to government intervention in the energy marketplace, far less eager that the failures and shortcomings of intervention should be so expertly held up to public view.

SUMMARY

In July, 1979, President Carter went to Camp David, Maryland, to ponder the content of his forthcoming energy message. The President summoned a multitude of advisers and counselors to worry with him over the ideas he would advance. No message of the administration's entire tenure was so thoroughly hashed over, so subjected to scrutiny from every angle.

The irony is that the matter is so transparently simple — at any rate, up to the point that it abuts on more complex issues like nuclear energy. There is one way and one way only to maximize production of domestic oil and gas and to quench, in the short run, the nation's energy thirst. That way is to remove from oil and gas the whole burdensome weight of government controls. Throw them off altogether and let the free market operate, as it operated with such dramatic success until 1954 and, with respect to crude oil, until 1971.

A recent study by one of the authors suggests that not even now is it too late to achieve energy independence. Oil and natural gas may be an exhaustible resource, but this hardly means they will cease to be found overnight. It does, in fact, mean that the oil and gas finding rate, in terms of barrels of oil-equivalent found per foot drilled, declines with time. Declines how fast? Historical evidence points to an annual decline of 3.4%.

Now what if the U.S. had taken the 1973/74 embargo seriously, and what if it had responded with a vigorous exploration and drilling program (as opposed to the heavy-handed regulatory program that was, in fact, imposed)? In this event, energy independence could have been achieved by 1980. Indeed, ignoring the formidable political obstacles now firmly in place, energy independence could still be achieved on purely technical grounds.

For example, if the current U.S. drilling effort were permitted to increase at 20% per year until four times today's drilling rate was achieved, energy independence would come by 1994; moreover, it could be sustained well into the first decade of the new millenium. But suppose the current drilling effort of 240 million feet per year is sustained, without further growth? Then our import dependence will become ever more frightening until, by the year 2,000, foreign oil and natural gas will represent two thirds of the total U.S. consumption.

The whole complex of government controls, regulations,[1] stipulations, specifications is worse than a failure; it is a fraud. Bad enough that the controls have not worked to encourage energy production; worse that they should be used to mask the government's lack of a coherent energy policy — and to persuade trusting citizens that the energy producers are the culprits in this whole affair. To blame energy producers for producing too little energy is, in itself, a curious idea. Is it that the producers are lazy? Then why do not more industrious producers arise to take their place? Is it, on the other hand, that circumstances beyond the producers' control inhibit their efforts? And if so, which circumstances and how can they be removed? Those who, for whatever reasons, espouse government control of economic processes cannot consent that such questions should be raised. They embarrass. They discomfit. Someone might get the idea that government had no idea what it was doing.

Of course, in a sense, the controllers know what they are doing. Their notion is to bring economic activity increasingly under the federal thumb so that it may better *serve the public purpose*, whatever such a term may mean. The controllers believe the public is best served by low prices, kept low by the intervention of the federal government. "Low" no longer means what it meant in the early '70s, yet still the controllers are loathe that the unregulated marketplace should have any say in determining prices. Large profits might accrue to the producers — obviously at the expense of consumers!

It sometimes seems that fear of profits for oil men is chiefly what hardens the hearts of the federal controllers. There is — sometimes beneath the surface of their rhetoric,

[1]For a discussion of how American consumers living east of the Rocky Mountains were deprived of relatively easy access to Prudhoe Bay crude oil through an impenetrable network of rules and regulations, see Appendix A at the end of the book.

sometimes on the surface itself — a pronounced dislike of oil companies. Pop psychoanalysis serves no immediate purpose. Why the petroleum industry should be singled out for dislike and opprobrium is a matter for the historians and social scientists to concern themselves with. At all events, oil companies are looked on with disdain. It cannot be conceded that the oilman is a man like any other; he has to be a deep-dyed villain with his hand perpetually in the public's purse. Of course he cannot be accorded the profits that his labor, energy and imagination would yield him were the free market allowed to function. The *government* must determine what is a fair price — fair to producers and consumers alike.

What an extraordinary thought! For centuries, where given latitude, the marketplace has determined the fairness or unfairness of a given price. A fair price is whatever the producer and consumer say it is — the one by offering his product on specified terms, the other by accepting it on those terms. There is no other rule of thumb in pricing. All the ponderous government pricing formulas must fall if producer and consumer do not agree between themselves as to what is fair.

It is a proposition so transcendently simple and clear — as though written in fire across the heavens — that the wonder is any other approach could ever be tried. But tried energy price controls have been. For 25 years they have been tried. In that period the nation has gone from energy surplus to energy shortage. A country splendidly rich in the resources that power machines, heat homes, and propel automobiles must import approximately half of its oil supplies — a sizable percentage of those imports coming from countries with less than friendly feelings toward the United States.

The government, to be sure, has lately moved nearer price decontrol than at any time since the Arab oil embargo. Controls on crude oil prices are to expire in 1981, controls on natural gas prices in 1985 — unless, Congress should intervene and stay the process. Yet the government's love of meddling is not so easily appeased. Gasoline price controls remain in place, though there is always talk of eliminating them too. No less unfortunate is the "windfall profits" tax enacted in early 1980. The passion for reducing oil company revenues, having yielded a little to common sense, has metamorphosed into a new passion — one for confiscating oil company profits outright and putting them to a different use entirely.[2] And to what end? Not the production of energy, surely.

For production, wholly different remedies are needed. Remove all economic controls. Take them up none too tenderly and throw them on the junk heap. When is the public interest served half so well as when buyers and sellers meet in a climate of freedom, deciding for themselves what they want? He who doubts whether the free market works has only to consider what copious energy supplies the climate of freedom once provided us — and how quickly government meddling caused them to vanish.

[2] For another — and different — view on oil politics, see Paul Craig Robert's article entitled "Oil Politics," reproduced from *The Wall Street Journal* as Appendix B.

APPENDIX-A
THE DEMISE OF THE CALIFORNIA-TEXAS CRUDE OIL PIPELINE PROJECT

In May, 1979, the Standard Oil Company (Ohio) abandoned its five-year effort to build a pipeline from Long Beach, California to Midland, Texas. The pipeline would have supplied the South and Midwest with 500,000 barrels a day of Alaskan North Slope crude oil. Whereby hangs a tale of confused and contradictory purposes, of missed opportunities, of high and noble purposes gone astray — a portrait, in miniature, of the energy crisis.

It was not that government at any or all levels purposely thwarted the project. It was that government policies previously decided on, previously implanted in concrete, worked against government's earnest desire to spread the supply of energy.

It came about in this manner:

Sohio, a major producer in Alaska's Prudhoe Bay oil field, planned its "Pactex" project with a view to making the most of a peculiar situation. With the completion of the Alyeska pipeline, carrying Alaskan crude to the "Lower 48," a glut of crude oil was projected to occur on the West Coast. Accordingly, Sohio conceived and planned the first pipeline to link the West Coast with the rest of the nation. Not only would the pipeline more evenly distribute Alaskan crude; it would increase the incentives for development of oil in Alaska and on the West Coast.

The project was a simple one — as pipeline projects go — involving a new, deep-water tanker terminal at the Port of Long Beach, and a pipeline some 1,000 miles long. The Texas end of the pipeline would tie into existing pipeline networks, thereby reaching refining centers on the Gulf Coast and in the Midwest.

The pipeline would have used about 800 miles of existing

lines, which would have been converted from natural gas to crude oil service. Only a little over 200 miles of new pipeline had therefore to be constructed. All very conventional — nothing extraordinary.

What was extraordinary, in the age of the energy crisis, was the opportunity to provision the country with oil from the biggest domestic field since World War II. The Prudhoe Bay field had been discovered in 1968; Sohio's leases there covered over 50 per cent, or about 5 billion barrels, of the supply. The Trans Alaska (Alyeska) pipeline was finished at last in 1977 after agonizing delays occasioned by environmentalists, fearful of damage to Alaska's ecology.

The earliest plans had indicated that most Alaska crude would be refined on the West Coast. But the OPEC embargo caused an immediate drop in demand for petroleum on the West Coast and likewise spurred plans to develop further California's oil reserves. In 1974, Sohio concluded that, within four years, Alaska's oil would leave West Coast oil supplies substantially in surplus. By Act of Congress, Alaskan oil could not be exported; to reach the tighter markets outside California, it would have to be shipped through the Panama Canal unless a new pipeline were laid, affording cheaper rates and greater security.

From Sohio's standpoint, the best route for a pipeline lay between San Pedro Bay, California and Midland, Texas. Not the least of the route's advantages was the speed with which the new line could be brought on stream — or so it seemed back in 1974. By 1979, the word "speed" would sound supremely ironic. Sohio had been forced by then to wade through a morass of more than 700 permits — state, federal, and local — required before construction could begin. In five years, before abandoning the project entirely, Sohio spent over $57 million not just planning the pipeline but trying to win approval for it.

In the dim past, concerns about any major project had expressed themselves in three questions: 1) Is the project technically sound? 2) Is it economically attractive? 3) Is adequate financing available? A fourth question, nagging and stentorian in tone, has arisen since the early '70s: Can the project be authorized in timely fashion?

The Pactex project met with ease the first three criteria. By the fourth it was destroyed.

The curious observer might speculate that the pipeline had been sabotaged by widespread opposition. Not so. Indeed, widespread opposition might have saved Sohio the $57 billion it sank into planning. The project was enthusiastically endorsed by all the relevant governmental officials and agencies, including the President, the Energy Department, and the governors of Texas, New Mexico, and Arizona. Gov. Edmund G. Brown of California endorsed Pactex provided air quality disputes among state agencies could be satisfactorily resolved.

Likewise there was grassroots support, as in Long Beach, which voted, 61 to 39 per cent, for letting the pipeline originate there. Even the energy coordinator for the Sierra Club, no sedulous friend of pipeline projects, testified at a Senate hearing that "Evaluating the Sohio proposal from a land-use standpoint, of all the terminal proposals, Long Beach is clearly preferable."

So what happened concerning the necessary permits? Sohio's problems fell essentially into four categories. First, procedural delays which stretched out the time for answering the company's petitions. A prime difficulty here was that many agencies could not issue permits until other agencies had acted. For instance, three different agencies — the Coastal Zone Commission, the Corps of Engineers, and the Environmental Protection Agency — required approval of all other permit applications before they would consider Sohio's application. It was a variation of Joseph Heller's classic "Catch 22." An EPA permit could not be obtained without a Corps of Engineers permit; a Corps of Engineers permit could not be procured without an EPA permit.

A second problem was that each agency's scope of responsibility was limited. Indeed, the objectives of some agencies were in conflict with the objectives of other agencies. Whereas the Coastal Zone Commission, concerned with land use, preferred that onshore receiving tanks for the Pactex project be located away from the docks, the air-control agencies involved wanted the tanks located adjacent to the docks, for the sake of reducing power needs and, therefore, air emissions.

Third, the permit-issuing agencies often required changes in the project that, when consummated, invalidated permits already granted. Thus the whole painful process would start over again.

And fourth, no system existed for promptly dealing with objections from the project's opponents. A long, meandering hike through the maze of the permit process revealed at every turn new opportunities for additional reviews and litigation. Pactex's environmental impact statement, shortly before Sohio scrubbed the venture, had been before the California Supreme Court for over a year, with no action taken. Indeed, after the long delay, the court merely remanded the case to the trial court level for additional hearings. And even if the report had been accepted, opportunities to challenge individual permits were manifold. Consider the difficulties in more detail.

By the end of 1976, Sohio, having selected the Long Beach-to-Midland route, had filed applications with the Bureau of Land Management and the Federal Power Commission. The Federal Draft Environmental Impact statement was finished; likewise California's environmental impact report. Sohio had also filed its air quality permit applications with the South Coast Air Quality Management District.

Yet disturbing and conflicting signals were proceeding from the California authorities. The chairmen of the state air resources board and the state energy commission announced opposition to the pipeline because of possible environmental problems in the Los Angeles Basin. Both suggested other areas as possible sites. Long Beach Harbor, however, remained the choice of other state officials. And so, despite two years of planning, designing, engineering, and permit applications, there was not even general agreement that the project should be located in Southern California — if anywhere in the state!

The air quality problems in smog-bound Los Angeles were admittedly thorny. Sohio believed that the issue could be resolved at Long Beach through "tradeoffs" and the use of appropriate technology. Still, air quality remained the major source of controversy surrounding Pactex.

As 1979 began, numerous hurdles remained, but four years of work had produced major progress. The Environmental Protection Agency had issued all necessary federal air quality permits. All remaining federal permits and approvals either had been issued or soon would be. Congress had in fact passed specific legislation expediting these documents. The other states involved had issued — or were prepared to issue promptly — all relevant approvals. The Port of Long Beach, by a large majority, had approved a lease to Sohio for the marine terminal. El Paso Natural Gas and Southern California Gas Co. had agreed to let Sohio use 800 miles of their existing pipeline system. Sohio had, moreover, committed itself to a precedent-setting emission trade-off package estimated to cost the company more than $80 million, at a net benefit to the air quality of the Los Angeles Basin.

What, then, remained to be done? California had yet to grant the vital air quality construction permit; nor had the relevant agencies approved the air quality trade-off package and an application to locate two essential oil tanks at the terminal facility. Several lawsuits challenging the project's environmental impact remained before the California courts, and several others were threatened.

Sohio's effort to resolve the air quality problem had been a sincere and energetic one. Under the federal Clean Air Act of 1970, further industrial growth is barred in areas like Los Angeles, unable to meet specified air quality standards. However, federal policy permits new growth where the resulting emissions can be offset by reducing emissions from existing plants. How to achieve such reductions? Sohio owned no other Los Angeles area facilities whose pollution it could reduce. Yet, at last, the company evolved a plan. It would pay for a sulfur dioxide scrubber and equipment to remove nitrogen oxides at a Southern California Edison Company power plant, at a cost of up to $80 million. Would this suffice? One agency called the solution unacceptable; another called it the only acceptable plan. From both, Sohio needed permits. The company was still embroiled in the dispute when, in 1979, it said "enough."

Ideally, Pactex should have been ready to operate when Prudhoe Bay production reached 1.2 million barrels a day in 1978. As it was not, the West Coast had to move its surplus oil through the Panama Canal. The matter was all the more urgent because the surplus — Pactex's whole reason for existence — was projected to end during the late 1980s; after which the pipeline's useful life would have been terminated. Sohio concluded that if the final roadblocks could not be cleared away by mid-1979 and construction promptly begun, Pactex would have to be scrapped.

The company's chairman, Alton W. Whitehouse, met with Gov. Brown and other key state executives, hoping they might help get the pipeline off dead center. Leaders of the California Assembly were likewise consulted and agreed to back legislation speeding up the decision-making process. But the legislation was inadequate for Sohio's purposes: It would not have cut through the final difficulties in obtaining permits; more important, it would not have eliminated further delays resulting from lawsuits. Sohio looked at the alternatives. Sadly, it shook its head. There was no recourse. Pactex was off.

To be sure, there were expressions of shock and dismay from officials at all levels of government. But there could be no going back. As Whitehouse plaintively expressed it, "The fact of the matter is that neither the government leaders nor Sohio can turn around the results of five years of delay which have substantially eroded the economic attractiveness of this project." The fault lay not with any single official or agency but rather with the system, the process: the bewildering thicket of conflicting requirements through which the hardiest businessman walks with trepidation. As one Sohio official puts it: "Unless the process is changed, no significant energy project is going to be completed in this country without some extraordinary effort to get through the permitting process. And this applies equally to private and public investment." The nature of the solution may not be clear just yet, but one thing, given the nation's energy dilemma, is abundantly plain: *Something* must be done.

APPENDIX-B

By Paul Craig Roberts[1]
— Oil Politics

Energy was never a problem for Americans until the government stuck its nose into it. The further in the nose went, the worse the problem became.

Of course, it is not the same problem for everyone. As the government knows, there's opportunity in adversity. The budget for the new Department of Energy is already equal to half of the after-tax profits of all the major oil companies combined, and it gets all that money without having to supply a single gallon of gasoline to the consumer.

For decades the oil industry managed to maintain the nominal price of gasoline constant, which meant that it fell in real terms. But gradually cheap energy was undermined by the government. In 1954 an old statute was reinterpreted so the government could regulate natural gas producers. Holding down the price reduced the supply and shifted a larger percentage of energy use to oil.

In the late 1960s the attack on cheap energy intensified. The depletion allowances were reduced, which lessened the accustomed profitability of the oil industry even as clean air standards forced coal users to switch to oil. Miles per gallon declined as safety and environmental regulations added weight and exhaust emission controls to cars. In 1971 the government slapped price controls on domestic production and, after the 1973 embargo, established a maze of regulations that subsidized the importation of foreign oil.

But the real blow came when the U.S. government helped the Arabs set up their oil cartel. Few people know this story because it is not in the government's interest to tell it, and the oil companies are intimidated.

It became clear to the oil companies that the OPEC governments — which had been gradually shoving them out of their equity in the oil that they found and developed —

[1] *The Wall Street Journal*, April 10, 1979. Editorial page.

intended to turn the companies into tax collectors by forming a cartel and driving up the price to foreigners. The companies appealed to the Justice Department for an antitrust waiver so they could present a common bargaining front to the Arabs. They got it, but it didn't do them any good. G.H.M. Schuler, director of European operations for Bunker-Hunt Oil Co., testified before a Senate subcommittee in 1974 that State Department officials undercut the oil companies' bargaining position by telling the Arabs that the companies did not have the support of the U.S. government.

Some idealists in government saw in oil a way of extracting more foreign aid than the congressional appropriations process would disgorge. They actually believed that the Arabs would use the money to underwrite the economic development of the entire Third World.

Other sectors of the bureaucracy went along with the scheme for their own reasons. Patriotic elements, worried about the decline in American resolve and fearful that congressional liberals would rather put more Americans on welfare than defend our interests abroad, saw in the scheme a way of providing the Shah and Saudis with money to beef-up anti-Communist forces in the Persian Gulf and to pay African governments to expel the Russians. Since humanitarians and cold warriors each had a stake in the cartel, there was no incentive for either to point out the U.S. government's role in setting it up.

Much of the story is lying there in the public record. "Oftentimes the representative of the United States," testified Mr. Schuler, "becomes the representative of the country to which he is accredited." The lesson, said Mr. Schuler, is that once the producing states "recognized that they were not going to be restrained by the joint action of the industry, and that governments were not going to support industry, then the only thing to hold them back was their own self-restraint, and I do not think that is a realistic expectation. So the U.S. government and the oil companies by retreating in the face of demands, created a momentum which continues to carry through."

But our government learned a different lesson from the

Arabs — how to turn the oil companies into tax collectors. The President is going to free the oil price so the government can tax the higher profits at higher rates. Some tongue-clickers are sure that talk about a windfall profits tax is just a political cover. But they badly underestimate the government's need for revenues so it can balance the budget without cutting into its spending programs.

Senator Long, oil friend or not, is nevertheless under pressure to come up with revenues for his colleagues to spend in order to fend off an attack on his own prerogatives. Big spenders have let him know that his Finance Committee cannot expect to continue handing out billions of dollars through "tax loopholes" when their spending programs are running dry. Sen. Long perceives that big spenders, cornered by the public's demand for a balanced budget, are dangerous animals. They will strip the Finance Committee of its hand-out power in order to replenish their own before they starve in their corner.

Freeing price but not profit is not decontrol. By skimming profits so deeply the government would strip the oil companies of the ability to establish a private energy industry on the basis of new technologies and new energy sources. That task would fall to the government.

The President deserves more credit as a good agent of the government than he is getting. If he pulls this one off, he will be taking two steps forward under the guise of one step backward.